Studies in Space Policy

Volume 31

Series Editor

European Space Policy Institute, Vienna, Austria

Edited by: European Space Policy Institute, Vienna, Austria
Director: Jean-Jacques Tortora

Editorial Advisory Board:
Marek Banaszkiewicz
Karel Dobeš
Genevieve Fioraso
Stefania Giannini
Gerd Gruppe
Max Kowatsch
Sergio Marchisio
Fritz Merkle
Margit Mischkulnig
Dominique Tilmans
Frits von Meijenfeldt
https://espi.or.at/about-us/governing-bodies

The use of outer space is of growing strategic and technological relevance. The development of robotic exploration to distant planets and bodies across the solar system, as well as pioneering human space exploration in earth orbit and of the moon, paved the way for ambitious long-term space exploration. Today, space exploration goes far beyond a merely technological endeavour, as its further development will have a tremendous social, cultural and economic impact. Space activities are entering an era in which contributions of the humanities — history, philosophy, anthropology —, the arts, and the social sciences — political science, economics, law — will become crucial for the future of space exploration. Space policy thus will gain in visibility and relevance. The series Studies in Space Policy shall become the European reference compilation edited by the leading institute in the field, the European Space Policy Institute. It will contain both monographs and collections dealing with their subjects in a transdisciplinary way.
The volumes of the series are single-blind peer-reviewed.

More information about this series at http://www.springer.com/series/8167

Annette Froehlich
Editor

Legal Aspects Around Satellite Constellations

Volume 2

Editor
Annette Froehlich ⓘ
European Space Policy Institute
Vienna, Austria

ISSN 1868-5307   ISSN 1868-5315 (electronic)
Studies in Space Policy
ISBN 978-3-030-71387-4   ISBN 978-3-030-71385-0 (eBook)
https://doi.org/10.1007/978-3-030-71385-0

© The Editor(s) (if applicable) and The Author(s), under exclusive license to Springer Nature Switzerland AG 2021
This work is subject to copyright. All rights are solely and exclusively licensed by the Publisher, whether the whole or part of the material is concerned, specifically the rights of translation, reprinting, reuse of illustrations, recitation, broadcasting, reproduction on microfilms or in any other physical way, and transmission or information storage and retrieval, electronic adaptation, computer software, or by similar or dissimilar methodology now known or hereafter developed.
The use of general descriptive names, registered names, trademarks, service marks, etc. in this publication does not imply, even in the absence of a specific statement, that such names are exempt from the relevant protective laws and regulations and therefore free for general use.
The publisher, the authors and the editors are safe to assume that the advice and information in this book are believed to be true and accurate at the date of publication. Neither the publisher nor the authors or the editors give a warranty, expressed or implied, with respect to the material contained herein or for any errors or omissions that may have been made. The publisher remains neutral with regard to jurisdictional claims in published maps and institutional affiliations.

This Springer imprint is published by the registered company Springer Nature Switzerland AG
The registered company address is: Gewerbestrasse 11, 6330 Cham, Switzerland

Preface

In recent years, space activities have continued to increase in number, scope, participation, and ambition as entry barriers to emerging and commercial space actors have continued to fall. New national and commercial space efforts are being directed at pressing issues of social, environmental, economic, and broad developmental importance, and satellite constellations are central to all these focus areas. Satellite constellations play this critical role because they support and enable advances in communication, navigation, disaster monitoring, Earth observation, and scientific activities, amongst others. Moreover, their role in supporting daily life on the ground (for example through the Global Positioning System) cannot be understated, and the promise of satellite mega-constellations in low Earth orbit to expand Internet access has garnered more public attention and enthusiasm for satellite constellations than ever before.

However, serious legal- and policy-related concerns remain unanswered in the face of the rapid growth in satellite constellations, especially given the size and number of mega-constellations currently being set up. As such, this publication, which follows on the well-received first volume of *Legal Aspects around Satellite Constellations* published in 2019, expands on the scope of its predecessor and brings together a diverse range of timely and valuable legal insights on topics from a variety of fields. The contributions, drawn from across the globe, also address critical aspects that take into account the needs and interests of the whole international community.

Given the urgency of responding to the legal questions concerning satellite constellations, these contributions and the novel approaches and perspectives they provide is certain to be of great interest to practitioners, academics, and students working in the space, legal, and related fields, given the vital contribution they make to filling the gap in the discourse on these issues.

March, 2021   Dr. Annette Froehlich
Seconded by German Aerospace Center (DLR)
Vienna, Austria

Contents

Small Satellite Constellations, Infrastructure Shift and Space Market Regulation 1
Lucien Rapp and Maria Topka

Facilitating Small Satellite Enterprise for Emerging Space Actors: Legal Obstacles and Opportunities 29
Michael Gould

Approaches to and Loci for Regulation of Large and Mega Satellite Constellations 47
Jack B. P. Davies and Jonathan Woodburn

Potential Antitrust and Competition Challenges of Satellite Constellations 83
Alice Rivière

The Designation of Satellite Constellations as Critical Space Infrastructure 103
John Tziouras

Satellite Constellations and the Sustainable Use of Outer Space 123
Gina Petrovici

Environmental Principles of Corporate Social Responsibility and Their Application to Satellite Mega-Constellations in Low Earth Orbit 143
S. Hadi Mahmoudi and Aishin Barabi

International Cooperation as an Essential Part of the Galileo Programme 161
Tugrul Cakir

Rise of Mega Constellations: A Case to Adapt Space Law Through the Law of the Sea 179
Lauryn Hallet

Small Satellite Constellations, Infrastructure Shift and Space Market Regulation

Lucien Rapp and Maria Topka

Abstract With the commissioning of the first constellations of hundreds or even thousands of small satellites, we are witnessing today an infrastructure shift. While it has not completely distanced the exploration and use of outer space from the dramatic geopolitical and military implications they once entailed for States, it is undoubtedly transforming this realm into a new economic frontier of competition, with its predominant players, this time being private profit-driven actors sensitive to market forces. As the exploitation of outer space becomes more economically viable, new commercial services should emerge through the deployment of SmallSat constellations and the provision of services by means thereof, creating a risk of increased dependency of the services consumed on earth on these new infrastructures. Therefore, new legal challenges pertaining to competition, foreign investment and the global economy as a whole do arise.

1 Introduction

With the deployment of the first constellations of hundreds thousands of small satellites in low orbit, is the world space industry to be gained over by the 'monopolization' movement that has been observed in many sectors in the United States, for the past few years? A few predominant operators, mainly American, would exercise exclusive control over critical infrastructures through which a large number of space services, data or commercial applications would transit. One must note that most of the same operators who launch these constellations and propose to operate them are today subject to a tightened vigilance by regulatory bodies, particularly in the United States, for the conditions under which they operate powerful digital platforms. This

---

L. Rapp (✉) · M. Topka
Chaire SIRIUS–IDETCOM, Université Toulouse-Capitole, Toulouse, France
e-mail: lucien.rapp@ut-capitole.fr
URL: https://www.chaire-sirius.eu

M. Topka
e-mail: maria.topka@ut-capitole.fr

© The Author(s), under exclusive license to Springer Nature Switzerland AG 2021
A. Froehlich (ed.), *Legal Aspects Around Satellite Constellations*,
Studies in Space Policy 31, https://doi.org/10.1007/978-3-030-71385-0_1

vigilance could go as far as a decision—or the threat of a decision—to dismantle them.

However, economic science shows that a monopoly position is not necessarily condemnable. First, because a monopoly can be "natural", as has been claimed for many years for telecommunications. Even today, the local loop is a de facto monopoly that it has been difficult to reduce, other than through the obligation to unbundle each consumer's service line. Also, because a monopoly can and must be regulated to limit its dangers. These dangers are well known, starting with the risk of confiscation of the famous monopoly rent, abuse of domination and even more, sterilization of any innovation.

Do we want, despite some positive externalities,

– the profits of the Internet via satellites to be captured by a handful of individuals;
– the countries with deficient terrestrial infrastructures to be placed under the control of a few large world groups, mainly American;
– the considerable progress made over the last fifty years in the field of satellite communications, earth observation, space imagery and their many commercial applications to be frozen with the commissioning of these constellations?

It is not, however, the one that attracts the attention of the international community. It is the issue of debris, often raised as the most important. It is true that the proliferation of debris is a cause for concern, to the point of raising the question of whether it will still be possible to launch and put satellites into orbit in the coming years. Nevertheless, its answer calls for an international management of space traffic that can only be effective if an international institution independent of States does it. Space activities, despite their development and their importance in today's economies, are the only ones that do not have a specialized organization within the United Nations system. Moreover, the creation of an international civil space organization refers to the current disorders of global governance. It can only be seriously considered at the end of an international conference, which is impossible to convene in the surrounding international context because of too great a divergence of interests among the nations of the space club. The difficulty is accentuated by the recent increase in the number of spacefaring nations.[1]

In the absence of a specialized international organization, we live on expedients and hopes. Expedients with the considerable work done within an expert body, the IADC and the consequent development of recommendations, more effective and better respected, it is assured, than legally binding provisions. However, they have the disadvantage of pushing international law further into soft law, taking it further away from the law of treaties, which has been its strength. There is hope, particularly that virtuous space legislation, such as French law (LOS), will eventually become a reference, if not an example to be followed, for all States seeking to adopt national

---

[1] Today of more than 80 (85), whereas they were only about thirty to invest in the space sector, at the turn of the century, in 2000.

regulations. From one convergence to another,[2] one can thus think that a common international rule, no doubt customary, will emerge. In the long term, it will logically impose the creation of an international authority to verify compliance or to adapt its contours, on the model of the ICAO or the International Seabed Authority.

Beyond these considerations that mobilize the international scientific community, one would vainly seek a reflection on the monopoly situations that threaten the world space industry and its future. It is as if most other nations had abandoned world leadership to the United States. Without apparent resistance, the latter judiciously seizes this opportunity by proposing a re-reading of international space law, essentially political, strongly inspired, in spite of the diplomatic language in which it is expressed, by the objective of protecting their national interests. It is the Artemis Agreements that propose to the States that wish to do so—there are nine Signatories as of November 2020—to rally the American banner around ten new principles.[3] This approach can be based on the fact that the major treaties establishing international space law were signed by a small number of States, at the heart of a political episode that is now over, the Cold War, when space activities were limited to the prudent exploration of outer space. Today, however, space is both a place of exploitation, a business activity, and a fast-growing market that better adapted rules must promote.

Certainly, several of the principles proposed by NASA suggest an intelligent and innovative adaptation of international space law in the direction of market law. And the very pragmatic approach that the American space agency favors, that of bilateral treaties favoring the beginning of groupings of states, opens up useful and constructive opportunities in terms of the global governance of the commercial applications of space, its resources and its technologies. They both justify the attention of the international community, which cannot ignore them or reject them outright. However, they could inspire a less US-centric and more distant approach to a desirable evolution of the international legal framework for the space industry in the interest of the international community.

Should the international treaties of the 1960s be discarded, in favor of principles that authorize, without conditions or reservations, the private appropriation of space resources, the unilateral definition of security zones around satellites the imposition of visions and legal interpretations that best serve one's interests instead of international cooperation?

Under these conditions, can we be satisfied with the silence of the European States, with the exception of Luxembourg and Italy, which joined the Artemis Agreements? Unless the silence of the other European States is loaded with ulterior motives:

---

[2] D. Alary, L. Rapp, The way forward to ICSO: an International Organization to handle a sustainable Space Traffic Management (Paper code: IAC-19,E3,4,11,x50161), in: 70th International Astronautical Congress (IAC-19), Washington, D.C., 2019.

[3] The Artemis Agreements are part of the Artemis program launched in 2017 by U.S. President Donald Trump. This program is articulated in several stages, the end-point of which should be the sending of astronauts on the lunar ground in 2024. Ultimately, the goal is to build a sustainable infrastructure on the Moon's surface in preparation for future missions to Mars.

- the anticipation of limited profitability of second-generation American constellations in the face of the success of the 5G terrestrial networks deployed in Europe;
- the gamble on the development of commercial applications stemming from intelligent processing of space-derived data as opposed to the space infrastructure that carries them;
- pressure from regulatory bodies, including American ones, against vertical integration strategies that are less and less compatible with market rules.

Following on from these introductory considerations (1), this chapter looks back at the recent evolution of the world space industry over the last ten years (2) to highlight a movement of decoupling of space infrastructures (3). One of the consequences of this decoupling is the growing need for regulation of these infrastructures, which are primarily of a private origin and nature, and more particularly, for a definition of their status with regard to the major principles of international space law (4). This regulation requires the definition of a legal framework for which a wide range of solutions can be put in place, notably those stemming from competition law (5). A conclusion proposes the course of action to be taken to achieve this (6).

2 The Miniaturization of Satellites: A Silent Revolution

The miniaturization of satellites operates a silent revolution in the global space industry, less in itself because it is not in the nature of a disruptive innovation,[4] than by its effects: it brings the world space industry into the era of mass production and thus the "trivialization" of space techniques at the service of many terrestrial activities.

The miniaturization of satellites is not a big deal, otherwise one could object that the first artificial satellite, launched on October 4, 1957, the Russian Sputnik1 satellite, was already a small satellite. It is more a matter of mass: with a lower mass than the satellites of the previous generation, small satellites certainly have more limited capacities, including a shorter lifetime, but they are also cheaper to produce and launch than larger satellites.

Small satellite class	Mass range (kg)
Mini-satellite	100–500
Microsatellite	10–100
Nanosatellite	1–10
Picosatellite	0.01–1
Femtosatellite	0.001–0.01

---

[4]Please see Victor Dos Santos Paulino and Adriana Martin *"Satellite Miniaturization: Are new entrants about to threaten existing space industry ?"*, SIRIUS, 2016.

Launched in constellations, they overcome the disadvantage of lower reliability by being redundant with other satellites, with resilience rates close to 100%. Built in large numbers, they allow the production of satellites in bulk, which facilitates maintenance operations by replacing faulty satellites. This is all the more efficient as their launches, which are less complex than those of large satellites, can be more frequent, most often in "rideshare" mode.

Positioned in close orbits, they have low latency and make possible Internet via satellite, thus meeting the expectations of more than half of the world's population, who do not have access to it. More difficult to jam, requiring a larger number of ground stations and mobilizing the skills of several industrial players, satellite constellations are less vulnerable than those of larger satellites are. They are likely to serve, better than they do, the expectations of defense and territorial protection; which may explain the growing interest they are arousing among armed forces and the military institution. They are called upon to play a major role in earth observation, environmental protection, and the surveillance of sensitive infrastructures, especially energy. In the long term, they will serve new clients from a number of sectors, who will see many opportunities in them: banking and finance in particular.

Benefiting greatly from advances in artificial intelligence, quantum computing and blockchain technology, the miniaturization of satellites introduces, more generally, a new culture and new industrial models. This is what we can call the "commoditization" of global space industry, which covers both the on-line production of space objects, at low cost, using or reusing products (materials, equipment, software) available "off the shelf" and the spread of the uses of space technologies in the economy and society. In so doing, it promotes the appropriation of space technologies by many sectors of terrestrial activities by forging new links between space operators and industrialists in the mechanical engineering, transport, energy, distribution and agricultural sectors. It thus opens up the market for space applications resulting from the exploitation of data of space origin, produced in large numbers, processed by the sophisticated means of machine learning and rapidly transformed into useful services for industry or commerce.

These advantages result in economies of scale, further reducing the costs associated with the production of these satellites and their launch. These developments have also prompted the emergence of cost-effective and higher-paced ways to deliver these satellites into orbit, such as the development of micro-launcher systems and spaceports (including sea launch pads) dedicated to their operation. The resulting lower entry barriers have, in turn, motivated the emergence of new space actors and facilitated local access to space. Portugal's proposed spaceport in the Azores archipelago is a shining example.

Miniaturization is thus significantly changing the economic dimension of the global market for space activities, by multiplying the number of satellites to be launched each year, by encouraging the emergence of new activities (the provision of in-orbit services for example) and by requiring the renewal of several other, more traditional activities (launch, space insurance in particular). It is not surprising that the commercial applications they enable are multiplying, from satellite communication to earth observation, including a wide range of commercial data services for agriculture, energy and transport.

Inspired by Skybox imaging, Planet labs or OmniEarth, many specialized companies have announced their intention to launch constellations of small satellites and enter this new market. Around the constellations themselves, many others, like TrustMe, Maxar Technologies, Northrop Grumman, Rocket Lab, HyImpulse Technologies, Rocket Factory, Isar Aerospace, Black Sky Imagery, Spacebits, Satellogic, Momentus or Preligens … have also rushed into the open breach, positioning themselves upstream on their components or downstream on their applications.

3 Small Satellite Constellations, Decoupling and Competition

The commissioning of the first constellations of hundreds of small satellites causes a double decoupling: a decoupling of the global space infrastructure (3.1) made possible by a decoupling of the global space industry, shifting from public to private and, without anticipating the following developments, from an international regime to a national regime, today essentially American (3.2). Contrary to popular belief, however, this double decoupling may not be synonymous with increased competition (3.3).

3.1 Decoupling of the Global Space Infrastructure: Space Infrastructures as Enablers of New Industrial Ecosystems and of New Markets in the Space Value Chain

Space-based activities are often described as infrastructure-based, described as "critical" because of its vulnerability. In fact, it is the increased dependence of many terrestrial activities on satellites and, more generally, on all the services now provided from space, that create this vulnerability and fully justifies this qualification.

In any event, these activities and services can only be carried out, provided if they are based on means constituted by space objects, whatever their size or capacity, whether they operate in isolation or within constellations. These space objects themselves require for their operation design and construction capabilities, facilities for their launch, authorizations for their operation, terrestrial relay or control equipment, means for processing and making available space data, commercial agreements, computer systems, etc. This is what space infrastructure is all about.

This infrastructure has long been under the direct or indirect control of States and their space agencies or international organizations, including international cooperatives, which operated it. The value chain of the global space industry was reduced limited to the activities of designing, building, launching and operating satellites, including the provision of imaging, positioning or observation services. For each segment of this value chain, there were only a small number of operators whose activities were linked, vertically and almost hierarchically, between a prime contractor and several layers of subcontractors at different levels. This industrial organization was all the more simple and easy to identify as the number of States in the space club was limited and the world industry had become highly concentrated, especially in the first decade of the 21st century. It described an industrial universe organized in sectors, of which one could list all the actors. These sectors were themselves part of national or regional programs, mainly defined and financed by States. Moreover, the latter were often present in the capital of the main industrial or commercial operators.

The miniaturization of satellites is likely to shatter this industrial organization, while the space club is growing significantly, from about thirty nations to more than eighty today. The industrial ecosystems that had prevailed until now are gradually being replaced by industrial ecosystems that are deployed around the infrastructure of the main constellations.

This new infrastructure has three characteristics that determine its specificity compared to the previous generation:

- it is designed to meet very terrestrial needs: those of emerging markets such as East and South Asia, those of additional broadband capacity in the era of 5G and the Internet of Things, and those still linked to the need for ubiquity (positioning, surveillance, communication), mobility (transport and logistics), and the mobilization of increasingly precise information in a global economy. In so doing, it widens the gap between the **near space** dedicated to terrestrial uses of space technologies (communication, earth observation, positioning) and the **deep space**, which remains an object of exploration; the former can be financed by private capital, the latter remains dominated by major public programs launched by States and conducted by their national agencies;
- it is a source of new needs in outer space itself, for the maintenance of constellations whose satellites have limited capacities and lifetimes, also referred to as "midstream" activities. It requires the mechanization of construction operations and the establishment of real production lines for small satellites. It presupposes their availability "in bulk" and their launch, in "rideshare". It calls for the multiplication of intervention operations and their perpetuation in space itself, in the form of dedicated services (in-orbit services). But it is also the source of new concerns, related to orbital congestion and the accumulation of debris in near space and the need for surveillance, if not joint space traffic management (SSA, STM), which do not yet exist and will have to be created;
- it produces space-derived data whose considerable mass and the possibility of processing it in large numbers with advances in artificial intelligence open up the promising market for many commercial applications of space technologies. These space infrastructures are thus evolving into enablers of industries and markets that rely on the data sets and imagery transmitted to provide applications spanning across very diverse sectors, such as navigation, the oil and gas industry,

communications, forestry, agriculture or even the military sector. As a result of the operation of such space infrastructure the markets for applications is becoming increasingly diversified.

The value chain has thus been considerably enriched. The number of satellite manufacturers has increased, while upstream many companies have emerged for the development of components (fuels) or equipment (articulated arms). We can no doubt anticipate that tomorrow's satellite assembly operations, stations and new platforms will take place in space and no longer on earth. Launch operations are no longer the prerogative of a few global operators. The number of the latter has increased significantly, with the development of vehicles or launch facilities adapted to the requirements of small satellite constellations. The vogue for "re-use" implies an improvement in materials or propulsion modes that favors the emergence of small companies. Obtaining the authorizations required for launch or in-orbit control has become a commercial matter, such as leasing frequencies or assigning filings with the ITU. The market for the maintenance of small satellite constellations is developing around a multiplicity of incoming operators who position themselves in it. The exploitation of space-derived data is emerging as an Eldorado whose mastery no longer only involves the mobilization of space resources: it is companies in the sectors of quantum computing, machine learning, big data, which are investing in it. Space activities are therefore no longer limited to specialized operators in sectors dedicated to the construction, launch or operation of satellites; they plunge deeply into many terrestrial activities, gradually blurring the boundary between space and terrestrial industries.

With the development of constellations of small satellites, the space infrastructure is diversifying beyond reason, branching out to excess, becoming commonplace, losing its specificity. But in doing so, it is also becoming dematerialized, privatized and internationalized, increasingly escaping the control of states. Perhaps tomorrow it will be totally externalized, spreading out in space itself, without any established link with States?

3.2 Decoupling of the Global Space Industry: Emergence of New Industry Trends, Innovative Business Models and Flexible Funding Schemes

Within a few years, several new companies have entered the global space industry. Observers generally focus on the successes of the American operator SpaceX (Starlink constellation), Amazon's projects (Kuiper constellation) or the problems with OneWeb, which was eventually taken over by the British government and the Indian operator Bharti Global.

But several other important companies were created in the wake of these companies which are today magnificent unicorns. On the American side, these include the operators Skybox Imaging, created in 2009, in which Google subsequently invested;

Planet Labs, created in 2010 (initially Cosmogia Inc.); NovaWurks, very active since 2012, before its takeover by Airspace Internet Exchange in 2019; OmniEarth and UrtheCast, both created in 2014; GeoOptics, Garvey Spacecraft Corp. and Silicon Labs.... On the European side, it is the operators NovaNano, Clyde Space, Gomspace, Deimos Space, Dauria Aerospace, Virgin Galactic (which, although incorporated in the USA, is a European conception), Swiss Space System, better known under the acronym "S3", CLS, Erems and many others, whose industrial initiatives are gradually taking shape with success.

Significantly, their projects have remained very attractive to investors in the year 2020. Despite the containment measures imposed by the COVID-19 pandemic, fundraising, which grew strongly in 2019, continued at a very good pace.

The new entrants, whether start-ups or experienced operators, all focus on exploiting the resources offered by the development of space technologies and the commercial opportunities they open up. For them, these techniques are not only a means; they have a commercial value. They place their projects or activities within the framework of vertical integration strategies, privileging the service dimension over the network and the logic of platforms, offering new forms of intermediation between service providers and customers.

The business models they propose are based on the new and very promising capabilities that can be offered by constellations of small satellites, mass-produced, which significantly reduces production costs, and positioned in close space, which limits the cost of launch. They also favor the choice of accessible technologies, often freely available. They are currently practicing the concept of Minimum Value Product (MVP), which allows them to reduce the Time to Market and to go from prototype to commercial product, testing its qualities and attractiveness to its future customers.

Finally, one of the success factors for these entrants is a new approach to the risks associated with space activities: namely, a considerably lowered risk aversion. Both in respect to the activities of space operations, de-dramatized and almost trivialized in the image of the space tourism project, and those inherent to the development of any economic activity, failure thus becoming an option—to use the reverse, one of NASA's famous slogans—that is integrated into the market approach and from which one hopes to recover quickly.

Beyond this first set of considerations, and undoubtedly because of them, what characterizes New Space is its ability to rapidly transform new technical concepts into commercial success. Hence the expression "Fast Space", which is also sometimes used. The table below lists nineteen commercial successes for five sets of traditional space activities, in the sense of companies that are now established in their respective markets. None of these companies existed more than ten years ago. They are all products of the first two decades of the 21st century and many of them were created in the years 2010. They are all or almost all American, which should not lead to exclude some other successes of the same type, Chinese, Indian, Japanese, Australian, Israeli or even European. There is no doubt that the table below can be completed and enriched with these non-American examples.

Business area	Technical concepts	Business concepts	Example companies
Communications	• Megaconstellations • LEO/MEO orbits • Smallsats and cubesats	• Vertical integration • $ per bit model • Direct-to-consumer	• OneWeb • SpaceX • Telesat • Leosat
Remote sensing	• Cubesats and smallsats • SAR payloads • High revisit rates	• GEOINT-as-a service • Value-add analytics	• Planet • Capella space • HawkEye • EarthNow • Orbital insight • Planet • Descrates labs
Weather	• Cubesats and smallsats • Multimode payload • High revisit rates • Precision weather	• Commercially provided weather data • Subscriptions	• GeoOptics • Spire • Orbital Macro System • ExAnalytic
Space situational awareness	• Onboard vs Off-board sensors • Multisource data integration • AI & ML	• SSA-as-a service • Subscriptions • Value-add analytics • Commercially-sourced data	• Applies defense • Agi
Launch	• Reusable launch vehicles • 3D printed components • Air- and sea-based launch concepts • Responsive launch	• Lower launch costs • More frequent launches • On-orbit delivery	• SpaceX • Vector • Blue origin

Credit Conference "Beyond the craddle 2019: envisioning a new space age", MIT 2019

The logic in which these activity projects or activities of a new type are inscribed is the antithesis of the public programs that characterized the beginnings of the space adventure and still mark, particularly in Europe, the development of its current activities. Because these companies cannot always count on public funding, they have built their projects on strategies that combine different factors:

– cost-efficiency—a particular and almost obsessive attention to costs and their reduction (e.g. developing reusability capacities, employing design-to-cost techniques, additive manufacturing),
– expanding one's customer bases—the generous promise of access for the greatest number of people to the new technologies resulting from advances in space research, echoing the founding principles of international space law,
– sustainable markets—an ecosystem-based approach to markets that emphasizes club effects and whose slogan is explicit: the more suppliers and customers there are in a given market, the more sustainable the market itself,
– new business trends and models—solid industrial partnerships that are only destined to become permanent as long as each partner remains the complement of the other and does not aspire to become its competitor,

– diversification strategies—vertical integration (into up-, mid- and downstream sectors), data-driven business models, careful gap analyses and attention to emerging market opportunities,
– novel financing methods and risk-sharing—since institutional funding is not always available, attention is turning towards private funding- and investment schemes (e.g. venture capital funds, private equity, seed capital, "angel investors"). As a result, risks and responsibility are shared with, or even transferred to, the industry rather than the public sector.

These strategies are not foreign to other strategies that have made the success of today's industrial giants, starting with Google. It is worth recalling that it is through an approach of the type just described that Google successfully entered the oligopoly of mobile operating systems, by making available an open source platform, itself built on a kernel of European origin (Linux).

3.3 Risks to Competition

The multiplication of constellation projects—more than 50, it is said at the beginning of the third decade of the 21st century—should not be abused. The markets, it opens up, are dominated by a few operators. They will no doubt remain so when, in a few years' time, an assessment is made of the projects imagined or initiated in 2020. The diversification of services or the attraction of private investors, in need of investments, in a financial climate that reduces their choices, will not prevent the movement of concentration that we have seen after liberalization, in the air transport or telecommunications sectors. And if the year 2020 was, despite the health context, an important year in the development of new companies and the appearance of new unicorns, like Momentus specialized in space transportation and valued at some 1.2 billion dollars for an order book of 90 million, it is marked by some initial operations of company mergers. Thus, AE Industrial Parners took control of Adcole Maryland Aerospace and Deep Space System. Redwire acquired Made in Space. Amergint Technologies Holdings acquired Raytheon Technologies. Amergint became the largest shareholder of Tethers Unlimited, just as Voyager Space Holding acquired Pioneer Astronautics.

It would certainly be wrong:

– to confuse the multiplication of constellation projects with a situation of full competition. Without anticipating the disappearance of several of them, it is highly likely that the market will gradually become hierarchical. Some of these projects, based on an architecture of hundreds or even thousands of satellites, which are generalist in terms of the services they offer, will dominate the market, offering a real private alternative to existing infrastructures, all of which are of public origin. The other projects are dedicated to uses that are too specific, if not captive in the sense of infrastructures reserved for the use of closed groups of users, to exercise market power;

- to exaggerate the depth of the new industrial field opened up by the miniaturization of satellites. No one can seriously maintain that the projects that are candidates for the now famous A, B, C, D+ series, and which sometimes raise significant funding (in addition to private equity), will all go through to completion. It is also impossible to ignore the fact that these projects are mainly American and that on other continents, start-ups are still multiplying in the traditional segments of the global space industry value chain (mainly the launch or construction of satellites), which are often losing growth today;
- to forget that the year 2020 will have been a special year for private investors, less because of the health context itself than because of its financial consequences. Central banks will have been substituted for private operators in the repurchase of public debt, even at the price of their independence. In an economic climate dominated by low rates, private savings, more abundant than ever because of containment decisions taken by governments or measures to support industry or trade, will have been redirected towards innovative projects, the miniaturization of satellites offering many opportunities, as described above. In many sectors of activity, stock market valuations will have soared, moving further away from economic realities, like that of the manufacturer Tesla, whose valuation now exceeds—by far—that of Toyota, whereas the former only produces at best a hundred thousand vehicles per year, while the latter produces 7 million.

Beyond these considerations, how can we fail to anticipate three emerging situations in the global space industry, which could ultimately constitute major concerns from a competition law perspective?

- Some of the most active operators in the small satellite constellation market today, such as SpaceX with its Starlink constellation, are vertically integrated operators. They are already present in several segments of the global space industry value-chain described above: the launch, satellite construction, operations and, probably in the future, data or services markets provided to the final consumer. They do not conceal the opportunity that the contracts they enter into with NASA represent for them: these markets offer them, in particular, the ability to launch their own satellites in the form of a rideshare. Can such behavior flourish without the interest of regulatory bodies, particularly in countries that are consumers of their services?
- Other operators, such as Amazon with its Kuiper constellation, are dominant in related markets, exploiting powerful electronic platforms, which are already the subject of investigations in many countries and at the center of a global debate, opened by the necessary evolution of competition law to take into account the many impacts of the digitization of the economy. In the audiovisual communication sector, there is an anti-concentration mechanism in many countries (please see here-below). It limits the market power of an operator who holds significant stakes in the press sector and wishes to invest in the audiovisual communication sector. There is currently no comparable mechanism for the space industry. Will this situation survive the commissioning of the first constellations of small satellites?

– Finally, among the planned projects, there are notably those of Russian or Chinese operators who are subordinate to their Governments or American or even European operators who are very close to their national agencies (SES) or even hosts public capital in their share capital (OneWeb). Can such a situation prosper without raising the question of the compatibility of these projects benefiting from significant public funding with the functioning of a free and competitive world market? In the wake of the Foreign Investment Filtering Regulation, European states are preparing to subject foreign investors benefiting from state aid to a reinforced control regime. Should not they also consider the future evolution of the market for small satellite constellations, to include them in the scope of their current concerns?

4 Small Satellite Constellations and International Law

At first glance, space law does not include provisions that could stem the competition threats they may catalyze. This regime was conceived in a context of Cold War ideological dichotomy and military antagonism; no commercial considerations fueled its institution starting with the COPUOS, at least apparently. The question may, thus, be raised whether the Outer Space Treaty ("OST"), which crystallizes the principles that are to guide space activities, could contain the emerging competition risks sufficiently. If not, then whether it could still serve as a point of reference to regulate these issues and thus, found a "space economic law", including rules or recommended practices applicable to the market that is opening up and to the private operators that are deploying in it.

The principle that is first enunciated in the Outer Space Treaty, is the **principle of free access to outer space** (Article I). Outer space is treated as a "province of all mankind" and, as such, all of its areas are free for exploration and use by all States, without discrimination, irrespective of a State's economic and scientific development, and for the benefit and interest of all countries.

All States enjoy the right to engage in space activities, including the launch and operation of satellites. However, the freedom of access to space or the right to non-discriminatory treatment are not explicitly granted to private entities. These rights are reserved to States only, which then are free to decide whether to extend them to their nationals or limit access to space solely to governmental institutions.

Certainly, we may be able to draw a parallel between the principle of free access to space and the competition law principles of market access and non-discrimination. However, it should be recalled that the drafters of the space law treaties did not envisage space as a market, but rather as a prospective Cold War front. Thus, these freedoms were not conceived in a market logic. After all, when the space law treaties were drawn up only governmental agencies would engage in space operations and their objectives were not commercial or profit-oriented.

In respect to the deployment of SmallSat constellations, several issues emerge. Orbital congestion is one of them: could the deployment of such constellations diminish the very access to space due to the sheer amount of satellites being placed into orbit? Orbital slots and electromagnetic frequencies are finite resources whose exclusive allocation to a limited number of actors would hinder the entrance of new competitors due to unavailability of slots and spectrum. This scarcity is currently at the origin of a new type of trade, that of filings with the ITU. Some of the constellations currently being deployed have benefited from these transactions, which raises delicate economic problems of valuation of these filings and some legal issues in the event of bankruptcy. Moreover, the now very sensitive issue of debris compounds this situation. By multiplying the constellations of small satellites, are we not running the risk of increasing the number of debris in nearby space and ultimately preventing the possibility of new launches?

That observed, the miniaturization of satellites is a factor in the downward trend in the cost of access to space. The construction of satellites has become inexpensive, satellites are mass-produced, launch operations are simplified with the possibility of rideshare, launch vehicles are recycled, launch operators have multiplied, and the number of spacefaring nations has increased. In fact, there may be a causal relation between the two phenomena of the multiplication of constellation projects and the increase in the number of spacefaring nations. By allowing more nations to gain access to space, the multiplication of constellations may actually be considered as serving the objective of the Space Treaty.

However, two issues remain that the Space Treaty does not address. The first is economic. Do commercial applications of space technologies require such a multiplication of constellations of small satellites? When this question is answered in the affirmative, anticipating that the more infrastructure there is, the more applications there will be, the question remains as to whether the first constellations of small satellites will not raffle off the whole market, according to the economic adage "the winner takes all"? The market could therefore concentrate and the number of constellation operators could therefore reduce, posing a second question: that of access to the infrastructure, which is a different issue from that of market access. Indeed, what happens to the principles of freedom and equality of access to space if the constellation market becomes oligopolized or even monopolized?

Further, the question of the status of the infrastructure could raise delicate international problems of direct or more likely indirect expropriation (by refusal or withdrawal of national licenses), or unilateral qualification by certain States or groups of States of the constellations as essential infrastructures, justifying ex ante regulation of their use. Foreign investment treaties will then have to take over the provisions of the OST. We can thus anticipate a multiplication of international disputes, mostly arbitral, in which the OST provisions and the principle of free access to space will likely be invoked.

The second principle crystallized in the Outer Space Treaty is the **non-appropriation principle**, according to which outer space is not to be appropriated by any State by claims of sovereignty, use, occupation or any other means (Article II).

The issue of non-appropriation is directly related to the term "province of all mankind" in Article I OST, which, however, should not be confused with the rather controversial notion "common heritage of mankind." The latter implies the existence of an international resource exploitation regime with benefit sharing and technology transfer among the parties, irrespective of their degree of involvement in the process.[5] Unsurprisingly, concerns have been raised that such a regime would entail economically detrimental, anti-competitive implications.[6] Instead, the "province of all mankind" principle may be analogous to res communis omnium, which implies that something is available for common use by all.[7] However, interpretations between developed and developing nations tend to collide in respect to whether this notion as implies limited or unlimited access to resources.[8] These divergences reflect a typical instance of the so-called "North-South divide", which could quickly become a "spatial divide" if we do not pay more attention to it. For reasons that can be explained with reference to the context in which it was signed, the Outer Space Treaty does not clarify how to handle these issues and their effects on international economy and competition.

Further, the exact meaning of "non-appropriation" also remains rather nebulous: Is it limited to the prohibition of territorial annexation of parts of space to a State, or does it go as far as to exclude private mining of outer space resources? Though we remain doubtful about the possibility of going as far as this exclusion, in any case, even if the non-appropriation principle allows for exclusive use of space resources, it does not set forth rules pertaining to their commercialization. This is a blank page that it is up to the States to draw up, either through their national legislations—organizing market access, setting the regime of access to essential infrastructures or sanctioning anti-competitive practices -, or in the form of multilateral treaties, even if these treaties are 'mini-treaties' on the model of the Artemis Agreements.

States are furthermore under **the obligation to carry out space activities in accordance with international law** (Article III OST) and **exclusively for peaceful purposes** (Article IV OST). It follows that international law is also applicable to the launch and operation of SmallSat constellations and the services provided by means thereof. As a result, even if the space law treaties do not contain explicit economic rules and antitrust principles per se, this provision may render these space objects and the services they facilitate mutatis mutandis subject to international economic and trade law principles.

Presently, there do not exist a unified or comprehensive international competition law regime or any international authority to enforce competition rules, which might be applicable to SmallSat constellations. Although national competition authorities often co-operate closely with their foreign counterparts,[9] competition law remains a

---

[5]F. von der Dunk, "International Space Law," *in* Fabio Tronchetti and F. von der Dunk (eds.), *Handbook of Space Law*, Cheltenham, UK and Northampton, MA, USA: Edward Elgar Publishing, 2015, pp. 57–58.

[6]F. Tronchetti, *Fundamentals of Space Law and Policy*, New York: Springer, 2013, pp. 13–14.

[7]F. von der Dunk, "International Space Law," *op. cit.*, pp. 57–58.

[8]J.L. Zell, "Putting a Mine on the Moon: Creating an International Authority to Regulate Mining Rights in Outer Space," *Minnesota Journal of International Law*, 2006, vol. 15, pp. 507–508.

[9]The 1991 EU-US Competition Cooperation Agreement constitutes such an example.

domain that is predominantly dictated on a national level. This results in disparities in competition rules and their enforcement. These issues may be further exacerbated by the extraterritorial application of a national antitrust legislation, easily leading to conflicts of jurisdiction.

As an alternative, recourse to the international trade regime developed within the bosom of the World Trade Organization ("WTO") may be possible, despite an international context in which multilateralism is being called into question and the authority of the WTO challenged. It must be noted that efforts to adopt a WTO agreement on competition policy have also been rather unfruitful.[10] Nevertheless, the services provided via small satellite constellations may fall within the scope of the 1994 General Agreement on Trade in Services ("GATS"). The GATS incorporates several competition-promoting clauses (e.g. national treatment, non-discriminatory market access, transparency) and requires States to eliminate barriers to the trade in services.[11] These services may also fall within the scope of the 1997 Fourth Protocol to the GATS (the "Telecommunications Agreement") which extended market access commitments to the telecommunications sector. Finally, the Agreement on Government Procurement ("GPA") may also be pertinent, particularly in respect to national preference and protectionism issues.

On the other hand, however, space-based systems and space-segments may also serve as critical infrastructure. As a result, competition in the space sector could be restricted on grounds of public order, security and military assessments.[12] The restrictions that the US Congress has imposed upon NASA, effectively banning it from cooperating in any way with China, constitute a pertinent example.[13]

Furthermore, States bear **international responsibility for their national activities in space**, including non-governmental ones (Article VI OST). However, since the Treaty is not addressed to private actors but solely to States, the burden of regulating private space activities is borne by the State. The latter must set forth an authorization and continuing supervision regime and ensure that these operations are carried out in line with the Treaty.

Apart from that, however, the increasing commercial uses of space also stress the urgent need for States to set forth measures guaranteeing competition and free market access in the space sector. Since the space law regime does not contain explicit competition provisions, the governance of space activities in this regard remains fragmented. The exact extent to which space actors will be required to refrain from anticompetitive conduct will depend on the national competition rules they are subject to.

---

[10] R.D. Anderson et al., "Competition policy, trade and the global economy: Existing WTO elements, commitments in regional trade agreements, current challenges and issues for reflection," *WTO Staff Working Papers ERSD-2018-12, WTO*, Oct. 2018, p. 59, available at: https://ssrn.com/abstract=3321116.

[11] See, in particular, Articles VIII, IX, XV, XVI and XVII of the GATS.

[12] F. Lyall and P.B. Larsen, *Space Law: A Treatise*, Surrey, UK and Virginia, US: Ashgate, 2009, p. 457.

[13] Section 1340(a) of *The Department of Defense and Full-Year Appropriations Act*, Public Law 112-10, 2011.

Furthermore, the principle of international responsibility begs the question whether a State could be held responsible for anticompetitive practices its nationals may have engaged in. After all, the space treaties do not specify what the term "national activities in outer space" encompasses. Of course, it is rather doubtful that the drafters of the Treaty had the provision of commercial services in mind when employing this term. Nevertheless, if it were to include commercial space activities, would a private company's anticompetitive conduct in a foreign market entail State international responsibility? Then what would ensue are probably questions of State jurisdictional immunity and non-justiciability before domestic courts, resulting in the need for international adjudication. Certainly, jurisdictional immunities can be waived in cases of commercial transaction (acts jure gestionis). However, would it be really justified to consider that the respective State virtually engaged in anticompetitive conduct instead and thus hold it responsible?

At the very least, we can expect some interesting arbitral awards on this subject, in the absence of jurisprudence from the ordinary courts. In this regard, it will be useful to check in particular whether the jurisprudence on the "abuse of automatic dominance" could be transposed to the constellations of small satellites. These are sanctions imposed on Governments or Agencies, which, by their administrative decisions (licenses, exclusivities, subsidies, public procurement ...), have contributed to the market power of certain operators and helped to build abusive dominant positions in their favor.

Further, Article VII OST stipulates that a State which launches or procures the launch of an object into outer space, or from whose territory or facility an object is launched, is **internationally liable for damage by such object to another State and its natural or juridical persons on the Earth, in air-space or in outer space**. Liability for damage caused by space objects is double-faceted: absolute liability applies for damage occurring on Earth, whereas damage occurring in space entails fault-based liability. However, in a context of proliferating small satellite constellations could the space liability framework be relied on to deal with damages arising from their utilization? It is far from being obvious.

Moreover, the space liability regime is riveted on the physical manifestations of the term "damage," namely loss of life, personal injury, and loss of or damage to property.[14] The advances in space technology, such as AI, commoditization of space data, the use of smart contracts and blockchain technology, may unveil new, non-physical manifestations of "damage", including damage resulting from anti-competitive practices, which the space liability regime is hardly keeping up with. However, such an evolution implies a wrenching revision of the approach to questions of liability related to space activities, which do not allow us to anticipate in the medium term an extension of the definition of damages covered, so that they include damages caused by anti-competitive behavior. Once again, national legislations only deal with these questions. These legislations are not specific to space activities and concern all sectors of activity; this may open the debate on their applicability to space activities.

---

[14] 1972 Liability Convention, Article I(a).

Further, the liability regime is intrinsically tied to the "launching State" notion. In the event, however, that the owner of a satellite is of different nationality than the entity providing services by means thereof, establishing responsibility for non-physical damage may prove to be a point of contention. Similarly, in the case of on-orbit assembled or 3D-printed infrastructure, defining the "launching State" may also be a challenge.

Moreover, even though the 1972 Liability Convention institutes a comprehensive dispute settlement mechanism for claims for compensation arising from space object induced (physical) damage, these claims can only be lodged by States against other States and not, against foreign private actors.

Tied with the concept of liability is also the **principle of registration of space objects**. Article VIII OST stipulates that **States retain jurisdiction and control over space objects they have registered**.

As space objects, SmallSat constellations will thus be subject to the legislation of their State of registry. However, competition rules and other market regulations set forth by the State it is being provided or its signal is being transmitted in, should also apply to the service provided by means of these satellite systems. This decoupling could gradually impose a legal characterization of the constellation as a space object from its status as an infrastructure used to provide a service, whose regime should be dealt with separately from that of the infrastructure itself. This "hybrid" legal status may raise questions as to which State will be held responsible in the event that a violation of the service-recipient State's competition laws occurs. Certainly, the OST does not provide a clear answer to this question.

Finally, the Treaty establishes the **principle of international cooperation in outer space** and requires States to ensure that their space activities comply with the principles of due regard (Article IX) and equality (Articles I and X). States are to disseminate information about their space activities (Article XI OST) and consult with other States if their space activity risks posing a potentially harmful interference to the latter's activities (Article IX OST). Despite these provisions, however, the Outer Space Treaty does not establish a specific cooperation mechanism or central authority to direct such cooperation.[15] Regardless, cooperation efforts are being undertaken on an intergovernmental and regional scale (e.g. ESA or the ISS cooperation). The promotion of cooperation efforts in the regulation of the commercialization of space-based activities would certainly be a very welcome development.

5 Small Satellite Constellations and Competition Law

Small satellite constellations are destined to play a global role. No one can deny that they are—and will be even more so as they come into service in the coming years—a

---

[15] S. Hobe and K.-W. Chen, "Legal status of outer space and celestial bodies," in R. S. Jakhu and P. S. Dempsey (eds.), *Routledge Handbook of Space Law*, Abingdon, Oxon and New York, NY: Routledge, 2017, pp. 35–37.

factor of progress. They will offer an alternative infrastructure to terrestrial networks and will promote connectivity. They will mobilize numerous activities around them and enable innovation. They will contribute to the development of information society services, starting with online commerce and transactions. The number of projects announced today suggests that their markets will remain competitive, even if some doubts can be expressed on this subject and a trend towards concentration can be anticipated.

However, they now have two drawbacks:

- they are not regulated, and for obvious reason, since they are not yet operational. However, they perform an intermediation function that may be at the origin of many behaviors that could be detrimental to their users, whatever their intermediary or final purpose: trade in personal data, illegally collected, processed or marketed, routing of illegal content, abuse of dominant position, anti-competitive or unfair practices ...
- when they are regulated, it is presently by American legislation, since most of them, due to an American technological advance, originate from the United States.

As it stands, international space law does not allow its provisions to regulate them. The five major treaties mentioned above were conceived at a time when the dominant concern was that of the exploration of space and its sanctuarization in a context of the Cold War. The modifications that are proposed within the framework of the Artemis Program, whatever their interest and that of the suggested approach are proposed by the American space agency.

It is therefore necessary to challenge this approach.

The constellations of small satellites and their main global operators are operating in a market logic. A solution could come from the implementation of new rules inspired by market law.

These rules relate to the control of concentrations, in particular through the implementation of an anti-concentration mechanism, such as the one that exists in certain States for audiovisual communication (5.1). Their purpose must be to monitor the risk of abuse of market position. Without going as far as dismantling de facto monopolies, which seems impossible in the state of the market for small satellite constellations, it would not be pointless to neutralize the infrastructure, avoiding the infrastructure operator from exercising a monopoly on the services to which a constellation gives access (5.2). It must be possible to apply the state aid regime to the operator of a constellation that has benefited from preferential market or public subsidy arrangements for the financing of its infrastructure (5.3).

5.1 Constellations and the Merger Regime

It is not so much the merger notification regime that we would like to insist on here as the specific so-called "anti-concentration" measures that exist in certain countries, such as France, to guarantee media pluralism. The first is applicable to operators in the space sector, whatever the constellation they operate or use. It is because of this principle that the Federal Trade Commission and the Department of Defense examined, in 2005, the grouping of Boeing and Lockheed Martin respectively within the United Launch Alliance (ULA). The European Commission's DG Comp. has similarly carefully considered, in 2016, the implications of Airbus Safran Launchers' acquisition of Arianespace on the launch market.

The second is more subtle. It aims to prevent the danger of dominant or even exclusive positions in the control of a global infrastructure through which transit services essential to the security of the serving country as well as to the functioning of its institutions, starting with its democracy.

Can the Canadian government conclude for some 600 million Canadian dollars, with the operator Telesat, an agreement to serve parts of its territory and to bring them the access to high speed Internet, while SpaceX obtains without conditions, an authorization to serve the Canadian territory for its Starlink constellation? Can the stated reason of fair competition between operators be satisfied? There is no doubt that each of the operators is within its rights and that additional competition in space can be beneficial to Canadian consumers. Nevertheless, should we not go beyond this first level of arguments, which are certainly admissible, to ask the question of the appropriateness of ex ante regulatory mechanisms designed to organize this competition, so that it remains truly fair. There are precedents here that could inspire the establishment of national mechanisms, on the condition, however, that all lessons are learned and that care is taken to avoid reproducing their complexity.

In order to guarantee media pluralism, which remains one of the conditions of the right to information and, consequently, of freedom of expression, the French law on freedom of communication, to cite just one example, includes a dual legal mechanism aimed at guaranteeing the internal pluralism of the media as well as their external pluralism.

The mechanism guaranteeing internal pluralism grounds on provisions limiting the share of share capital that the same individual or legal entity may hold in private audiovisual communication companies. The idea is obviously to avoid, if not prevent, the same operator from exercising too much influence on the information media. In French law, the thresholds are different depending on whether the capital concerned is national, European or foreign and they take into account the audience of each media concerned. Thus, Article 39 of the French Communications Act specifies: "the same individual or legal entity (...) may not hold, directly or indirectly, more than 49% of the capital or voting rights of a company holding an authorization for a national terrestrial television service" in analog or digital mode, "whose average annual audience (...) exceeds 2.5% of the total audience for television services". In the case of a satellite television service, the same article sets the threshold at "half of the capital or voting rights of a company" operating a satellite television service. Article 40 of the law of September 30, 1986 limits the participation of foreign persons outside the European Union to "20% of the share capital or voting rights".

In addition, a mechanism for controlling external pluralism takes into account whether or not the operator intervenes on one or more media. Thus, the law sets a maximum number of authorizations that the same individual or legal entity may hold for a media, as is also provided for the mandates of independent directors in a commercial company. For example, in terms of terrestrial television, no person may hold more than two authorizations if the cumulative potential audience of the services in question exceeds 20% of the cumulative potential terrestrial audience of all authorized television services. For services broadcast by satellite, the number of authorizations that may be granted to the same person for the operation of one of these services is limited to two.

An anti-media concentration mechanism complements this mechanism designed to ensure external pluralism. It allows the regulator to exercise control over the accumulation by the same person of authorizations held in several media (terrestrial television services, radio services, national or regional daily press), according to a procedure for assessing their market power taking into account; for example, a criterion of cumulative potential audience as described above.

Moreover, within the context of access to internet provided via SmallSat mega-constellations, net neutrality issues will also need to be considered. Of particular concern would be the relationship between the internet service provider and the providers of content. Vertical mergers between such actors would accentuate these concerns, particularly if such mergers lead to the employment of competition-restrictive tactics that would render competitor access prohibitive, such as bandwidth throttling, restricting IP interconnection capacity, prioritization of one's own services, limiting access or restricting directly or indirectly the possibility of content providers to distribute their content, etc.

Drawing a parallel with the EU'S TSM Regulation (as amended),[16] providers of internet access services and other electronic communications should be obliged to safeguard open internet access, for example by ensuring sufficient network capacity for the provision of high quality non-discriminatory internet access services.[17] They should treat all traffic equally, without discrimination, restriction or interference and independently of its content, application or service.[18] Considering, thus, that the actors aiming to provide internet access via satellite mega-constellations tend to be vertically integrated (construction of satellites, launch services, internet access services), further downstream integration into the content provider industry (e.g. in the form of vertical mergers) would not be unlikely. As such net neutrality rules will also need to be transposed and applied in respect these actors.

It is undoubtedly in the latter direction that a system applicable to constellations of small satellites must be devised, where the operator, who is necessarily foreign in relation to the countries served, cannot obviously be subject to a limitation on the financial holdings of its own shareholders. Neither is it necessary to set up companies under local law to make its infrastructure available and/or to provide its services.

---

[16] Regulation (EU) 2015/2120 of the European Parliament and of the Council of 25 November 2015 laying down measures concerning open internet access and amending Directive 2002/22/EC on universal service and users' rights relating to electronic communications networks and services and Regulation (EU) No 531/2012 on roaming on public mobile communications networks within the Union.

[17] See *TSM Regulation*, Recital 19.

[18] See *TSM Regulation*, Recital 8.

Once again, it is not a question of reproducing the complexity of these mechanisms, but only of laying down rules that are intelligible and useful, felt as such by all the operators concerned, simple to use and, consequently, likely to be applicable. Setting them at the European level to ensure an efficient regulation of the European territory, respecting the sovereignty of the Member States of the European Union, seems to be an indispensable precaution.

5.2 Constellations and Regime of Dominant Positions

The operator of a constellation of small satellites is in a dominant, if not exclusive, position in the provision of its infrastructure. Competition can only come from other constellations that are likely to provide the same type of service(s) from both a physical and geographical standpoint, or from terrestrial networks, cable networks or 5G mobile networks, provided that these terrestrial networks can be considered as substitutes for the small satellite constellation.

However, the possibility cannot be ruled out that:

- the market for connectivity will become more concentrated, that
- the number of small satellite constellations will be reduced, and that
- links will be established between terrestrial networks and satellite networks, through the provision of frequencies or capacity, exclusive agreements, or the exchange of shares, which could go as far as the takeover of a terrestrial operator by the operator of a small satellite constellation.

The hypothesis that the operator of a constellation of small satellites behaves like a digital platform and exercises access control cannot therefore be ruled out. It then becomes a gatekeeper, inclined to abuse its dominant position.

The European Commission recently reacted against this type of anti-competitive behavior by proposing two founding texts: a Digital Services Act and a Digital Market Act, of which drafts were made public at the end of 2020. These texts aim at the identification of practices by digital platform operators or online intermediation service providers, which are likely to constitute abuses of a dominant position.

These practices include, in particular, exclusive control of data collected for access to their infrastructures, preferential access rights for services promoted by these operators and suppliers to their infrastructures, and the marketing of applications provided by these operators and suppliers to the detriment of applications of companies using these infrastructures. The future European regulation focuses on ensuring trust and safety online by increasing responsibilities, obligations, and liabilities for digital services. It proposes also the establishment of an ex ante regulatory instrument to control the behavior of gatekeepers. This instrument includes prohibitions or restrictions of certain business practices that the Commission believes should be "blacklisted", reporting procedures, involving a new category of whistle-blowers ("trusted flaggers"), and obligations on platforms to modify their business practices to facilitate competition against themselves and remedies that would be applied on a case-by-case basis to large on-line platforms.

It is singular to note that these texts are aimed at terrestrial digital platform operators and online service providers; and some of them, qualified as operators of "structuring infrastructures". They do not explicitly refer to the constellations of small satellites, which may incur the same reproaches and justify the same vigilance. However, perhaps the final version of the two texts in preparation will reserve the pleasant surprise of definitions allowing the inclusion in their scope of all the digital infrastructures concerned, including space infrastructures.

This inclusion is even more necessary since some of the operators of small satellite constellations are vertically integrated. They operate in several segments of the value chain: construction and launch, launch and provision of connectivity services, provision of connectivity services, and provision of content or application services to access them. As a result, they hold market power that can encourage anti-competitive behavior.

The means available to the country of service include, in particular:

- the qualification of the infrastructure itself as an essential facility;
- the establishment of a right of access for third parties, which can itself be part of a policy of unbundling the infrastructure and the services it carries;
- within the logic of unbundling, the establishment of an independent regulatory authority charged with intervening on an ex-ante basis to ensure access to the satellite infrastructure and non-discriminatory entry in the satellite services market;
- the development of competing terrestrial or even satellite networks;
- the correction of market asymmetries by ex-ante regulation of the data or applications market;
- the obligation to ensure the interoperability of access equipment to these infrastructures.

Essentially, one of the main questions that will arise in this respect is whether regulators should promote a facilities-based competition or instead designate satellite infrastructures as essential facilities and require the infrastructure-holders to grant access to them (e.g. via compulsory licensing). This decision will require several considerations, such as for example whether the input in question is essential to enter a market, or whether there are viable alternatives or substitutes for the infrastructure in question. If lack of such input is likely to preclude competitive entry, then the facility is more prone to be considered essential and vice versa. It should be noted that even intangible assets, such as information and data collected via satellite infrastructure, may even constitute essential facilities.

We should highlight, however, that there are several constraints that ought to be taken into consideration when applying the means presented in the list above, as well.

On the one hand, infrastructures deemed, as "essential facilities" must allow third party access and entail the risk of allowing their holders to control competitors' access to a market. If unjustified or unreasonable, such denial of access to the infrastructure would indeed constitute an abuse of dominant position. On the other hand, however, the existence of objective justifications of denial need may plausibly include the non-availability of excess capacities, technical obstacles rendering access impossible, the existence of economically viable alternatives, or even the competitor's ability to provide reasonable remuneration.

Furthermore, will a facilities-based competition (in the form of requiring competitors to develop their own mega-constellations) actually increase efficiency? Alternatively, will it in reality reduce incentives to invest due to prohibitive market-entry costs? Additionally, recalling the issue of orbital congestion that is bound to be exacerbated with the multiplication of constellations, it might in fact not be sustainable to promote a facilities-based competition in the long run considering the limited availabilities of slots and frequencies.

Similarly, the existence of monopolies should not be interpreted as automatically implying abuse of dominant position, particularly when they do not impose unfair trading practices or prejudice consumers in other ways. Additionally, article 106(2) TFEU could also serve as reference in this respect since it states that undertakings performing services of general economic interest are subject to the rules on competition as far as the application of such rules does not obstruct the performance of the particular tasks assigned to them. Could the provision of internet access to remote areas be counted among these services? Regardless, care must also be exercised so as to avoid instances of regulatory capture that would facilitate abuse of automatic dominance.

Overall, dominant positions may reflect efficiencies and their regulation will need to balance free competition with the interest of safeguarding sufficient investments in such infrastructure.

5.3 Constellations and State Aid Regime

Some of the announced small satellite constellations benefit from the support of their governments or national space agencies, which directly finance their space industry through public subsidies or help their deployment through a very aggressive public procurement policy. These subsidies or contracts may distort competition in the markets for the provision of connectivity services.

This is a more general issue of interest in international trade law.[19] It has determined the European Commission to publish on June 16, 2020, a White Paper on establishing a level playing field for foreign subsidies.[20]

This White Paper proposes the establishment of new legal instruments, in the form of three modules covering respectively:

- the measurement of the effect of distortion of competition on the European Union market of foreign companies receiving subsidies or contracts benefiting them,
- the control of the acquisition by such foreign companies receiving subsidies or contracts, of European companies in the context of foreign direct investment operations, and,
- restrictions on access by these foreign companies to the market procedures launched by European contracting authorities.

These legal instruments must be applied to foreign operators of small satellite constellations wishing to provide connectivity services to European users both in their practices on the European market and in their investment operations in European companies; or even in their application for competitive tendering procedures on European soil. In this regard, particular attention should be paid to foreign investors, in particular State-owned or controlled ones, seeking to acquire, control or influence European undertakings carrying out activities that have significant spillovers on the European economy and society. That should be activities in sectors involving key assets and critical infrastructure (whether it is tangible facilities, such as networks, or intangible ones, such as information, technologies and other kinds of inputs). Certainly, the operation of SmallSat constellations could be comprised among them. The degree of operational and national security risk, such as access to sensitive information concerning critical infrastructure, will definitely have to be considered in those instances.

It should be noted in this respect that, while foreign State-funded acquisitions of such European assets are numerous, European companies cannot receive similar subsidies or support from their domestic governments due State aid rules. In addition, they do not enjoy similar acquisition or control opportunities in foreign markets, either. In fact, they might even face discrimination and differentiated treatment in foreign procurement bids. When competing in the American market for example, they have to deal with regulations such as the "Buy American Act" that requires the U.S. government to prefer U.S.-made products in its purchases.

---

[19] This issue is covered by the Agreement on Subsidies and Countervailing Measures concluded on April 15, 1994 under the Marrakech Agreement establishing the World Trade Organization (Annex 1A, 1869, U.N.T.S.14). However, certain OECD reports indicate that despite this agreement and the obligations it imposes (in particular the obligation to notify these subsidies and countervailing measures), these subsidies continue to grow, sometimes reaching very significant amounts (between US$20 and US$70 billion in the aluminum sector between 2013 and 2017). Moreover, the prohibited subsidies and countervailing measures only concern imports of products from third countries; in the case of the European Union, they leave out of their scope the intra-Community trade in services provided by foreign companies established in the territory of the Union.

[20] 17 June 2020, COM (2020) 253 final.

The European space launch industry specifically has been particularly impacted due to foreign competitors operating in such institutionally "captive markets" that favor domestic operators. For example, American, Chinese, Japanese, Russian and Indian launch service providers can benefit from exclusive access to their governmental markets, "generous" institutional contracts and long-term procurement contracts, which, in general, is not the case for the European market, putting European operators such as Arianespace at a competitive disadvantage compared to their foreign counterparts.[21] The question may thus be raised whether the EU should promote similar national preference rules when it comes to such key sectors.

Based on these considerations and in light of these challenges, the other side of the "State-aid coin" must thus be also considered by the EU: State aid and government procurement can serve as a stimulus to foster competition against existing monopolies and oligopolies. For example, the success of SpaceX, which has evolved into a fierce competitor in the launch service provider industry, might not had been effected had NASA not encouraged entry into the launch sector by SpaceX and other private companies with its Commercial Orbital Transportation Services program. Relaxation or at least a more flexible application of State-aid rules must definitely be enacted for such a strategic sector as space and the satellite industry.

From an EU law point of view, the public funding of such ventures, such as the development of satellite constellations and satellite-based services, may be justified in light of Article 107(3) TFEU. At least, as far as it concerns, for instance,

- "aid to promote the execution of an important project of common European interest," or
- "aid to facilitate the development of certain economic activities," or even
- "aid to promote the economic development of areas where the standard of living is abnormally low or where there is serious underemployment" such as the provision of internet and other satellite services to very remote areas.

These provisions relate also to the principle of "juste retour" applied by the European Space Agency: under this approach, domestic space industries are granted ESA contracts based on their respective State's financial contributions to ESA programs.[22] This approach may, actually, be considered as an indirect State aid, considering that contractors are not chosen on competitive grounds (e.g. value for money) but simply on grounds of their nationality.[23] Nevertheless, this practice has been exempted from the EU Commission's State aid rules due to the special, highly strategic, and highly

---

[21] ASD—Eurospace, *Aggregation of European institutional launch service levels*, Position Paper, 2018, available at: https://eurospace.org/wp-content/uploads/2018/07/eurospace-pp-on-aggregation-of-european-institutional-launch-services_july-2018.pdf.

[22] F. von der Dunk, "International organizations in space law," in Fabio Tronchetti and F. von der Dunk (eds.), *Handbook of Space Law*, Cheltenham, UK and Northampton, MA, USA: Edward Elgar Publishing, 2015, p. 312.

[23] F. von der Dunk, "European space law," in Fabio Tronchetti and F. von der Dunk (eds.), *Handbook of Space Law*, Cheltenham, UK and Northampton, MA, USA: Edward Elgar Publishing, 2015, pp. 265-266.

cost-intensive nature of the space industry and the fact that it serves overarching European interests.[24]

These considerations demonstrate the need to apply European State aid rules in a more flexible manner so as not to hamper the growth over European space operators.

6 Conclusion: Towards a Space Market Act

It is therefore desirable that a body of specific international rules be gradually put in place to organize competition on the emerging space activities market. This body of rules could consist of ten or so principles including a list of potentially unfair practices that could be committed by operators of small satellite constellations following the model of the list drawn up by DG COMP in the preparation of the Digital Market Act on anti-competitive practices of so named terrestrial "Gatekeepers Platforms".[25]

The forthcoming entry into service of the first constellations of small satellites could provide an opportunity to do so. Their activities are part of a major change in space infrastructure and global space industry. However, most of these constellations are under American jurisdiction. And we could expect that the U.S. Government and its agencies, including the market regulatory bodies in charge of antitrust in the United States, have no interest in proposing rules that could appear to their operators as constraints incompatible with their international development.

The initiative must therefore come from the European Union. On the model of the Digital Market Act that the European Union is about to adopt, one could easily imagine a Space Market Act. This Space Market Act could just as easily be an appendix to the Digital Market Act, unless it is included in its provisions.

The Space Market Act would mark the will of the European States not to allow private constellations of small satellites to replace yesterday's space infrastructures and to provide their commercial services, without a regime that respects both elementary rules of competition and the universal principles that the Space Treaty has laid down since the beginning of the conquest of space.

In the same way that NASA conceived and proposed the Artemis Agreements in mid-summer 2020, the European Union could suggest to the other spacefaring nations to join them by expressing their adherence to the Space Market Act or by concluding bilateral treaties with it.

Better yet, each State-member of European Union—including Luxembourg and Italy, which have already joined, but could express European solidarity here—could make its adherence to the Artemis Agreements conditional on the reciprocal adherence of the US Government and US agencies to the Space Market Act. However, the Artemis Accords are open to signature to individual States only, and not to international organizations (as per Section 13(3) of the Accords), what does in fact illuminate certain challenges and constraints that the European states would have to overcome first prior to being able to adopt this approach and promote a Space Market Act.

These challenges and constraints are essentially fourfold:

---

[24] Ibid.
[25] DG CNECT/GROW Informal Working Document, DG COMP, 18 September 2020.

- The Union will have to draw up a European space policy that incorporates elements of competition law or requires the application of competition law to space activities, with the aim to establish a unified European approach in respect to the commercial and competition concerns relating to space matters.
- The Member States, in turn, would have to coordinate with each other and apply the rules set forth individually as well, which however could prove to be problematic considering that space is not an exclusive competence of the EU. In addition, Article 189(2) TFEU excludes any harmonization of the laws and regulations of the Member States in respect to measures taken to promote the objectives of a European space policy.
- Moreover, convergence between the legal orders of the EU and ESA will also be required, so as to strengthen a European stance even further. The lack of a unified European approach is magnified by the fact that ESA and the EU are two separate institutions, the one not subject to the legal order of the other, and not all ESA member States are EU members and vice versa. This will entail challenges in respect to the promotion of the EU and ESA's objectives and the application of EU law to space activities or even the promotion of a European space economic diplomacy. So, for the EU and ESA to promote a Space Market Act that could pave the way for a future international space economic law, these issues will need to be dealt with first.
- Overall, the aim will not be to create a legal and normative Cerberus that would overregulate the operation of satellite constellations or the provision of services by means thereof, but to institute a competitive, transparent and efficient international environment that will promote innovative uses of small satellite constellations, with the end-goal being the maximization of consumer welfare and economic growth.

Such an initiative should not be incompatible with a more offensive policy, based on the mobilization of European industrial capacities, in a sector where several European States have historically occupied a prominent position. Yet, the European Union has decided to lead the charge in the field of space activities, by launching the project of a European constellation of satellites competing with constellations Starlink and Kuiper. May we hope in this way that, thanks to the legal and economic challenges posed by the commissioning of the first American mega constellations, space Europe could finally be back?

Lucien Rapp lectures at the Law School of the University Toulouse-Capitole (France) as well as at HEC, Paris (tax and law department). He teaches international business law and competition law. At the University Toulouse-Capitole, he heads the Chaire SIRIUS (Space Institute for Research on Innovative Uses of Satellites), an academic chair created in 2013 by CNES, Airbus Defense and Space (ADS) and Thales Alenia Space (TAS).

Maria Topka is a research associate to the SIRIUS Chair.

Facilitating Small Satellite Enterprise for Emerging Space Actors: Legal Obstacles and Opportunities

Michael Gould

Abstract The recent adoption of small satellite technologies by private space actors, due to relatively low costs for their development, launch and operation, has led commentators to point to their emergence as a means for the diversification of space actors. In essence, these constellations present a rare opportunity for international start-ups to bridge the perceived 'space gap', compete with the dominant commercial space oligopolies and better fulfil the egalitarian principles of the Outer Space Treaty. Unfortunately, this technological opportunity is being impeded by a lack of evolution in the legal framework governing space activities, and this paper seeks to identify three central tenets of legal impedance for small satellite enterprises: (1) textual uncertainty in treaty law (2) discriminatory liability requirements and (3) a lack of vision for space debris remediation. It will be submitted that a binding addition to public international space law, one that resolves these problems for smaller satellite technologies, is necessary to unlock the benefits of satellite constellations for developing and developed space nations alike and enhance innovation in the commercial space industry.

1 Introduction

Space technologies have provided immeasurable benefits to the societies found on Earth. Immediately tangible benefits include global communication, enhanced navigation and better meteorological understanding; however, space technologies have also provided intangible benefits such as better international cooperation, industrial organisation, more rapid technological development and personal well-being improvements.[1]

Nonetheless, there has emerged what is known as a 'space gap' whereby developing states have limited access to those technological opportunities gained through

---

[1] Doyle S, 'Benefits to Society from Space Exploration and Use' (1989) 19 Acta Astronautica 9, 750.

M. Gould (✉)
University of Bristol, Bristol, UK

space activities or must pay exorbitant amounts of money to foreign governments for basic levels of access.[2] This gap has contributed to the production and maintenance of outer space monopolies and oligopolies and renders space an unfortunate contender for the same imperialistic domination that preceded the space-age. Seemingly, the only way to challenge the domination of the major space actors is for emerging space actors to advance their own sophisticated space technologies, and small satellites (and the constellations they create) may provide this essential ingredient to invigorate market competition in space due to their comparatively inexpensive production. Small satellite enterprise could have both an economy-stimulating and economy-strengthening role in developing markets, democratising the commercial use of outer space and fulfilling the egalitarian paradigm envisaged by the legislators of the 1960s which saw outer space as a province that should be preserved for the 'interests of all mankind'.[3]

Unfortunately, public international space law does little to accommodate the economic liberation of the prospective space actors via small satellite enterprise. Substantively, the law surrounding small satellite activities is uncertain and has a disincentivising effect on prospective commercial actors. Further, space law renders implausible space debris remediation which threatens to impede all future space access if left unresolved. This paper will conclude that, although the legal regime may *de facto* apply to small satellites in constellation, new 'hard law' must be introduced to facilitate their use as a means of democratisation. If not, space law merely serves to preserve the ability of private space oligopolies to dominate space markets through a competitive mastery of this global commons,[4] a scenario which will reduce the market diversification necessary to stimulate global innovation within the international community so that all persons on Earth might benefit from satellite constellation technology.

2 Small Satellites and Space Law

2.1 Small Satellite Enterprise

Small satellites are part of the 'NewSpace' economy that favours initiatives led by private funding and business as opposed to the traditional model of space enterprise

---

[2]Papparlardo J, *'A Satellite For (and from) Africa'* (2017) Air and Space Magazine, https://www.airspacemag.com/not-categorized/a-satellite-for-and-from-africa-17768287. Last accessed 10th March 2020.

[3]Treaty on the Principles Governing the Activities of States in the Exploration and Use of Outer Space, Including the Moon and Other Celestial Bodies (Jan. 27, 1967) 610 U.N.T.S 205 [herein the Outer Space Treaty]; *also see* Way T, *'The Space Gap, Access to Technology, and the Perpetuation of Poverty'* (2018) 5 International ResearchScape Journal 7, 1–20.

[4]Bormann N and Sheehan M, 'Securing Outer Space: International Relations Theory and The Politics of Space' (1st edn, Routledge 2009), 43.

via governmental agencies.[5] These types of satellite include 'CubeSats' and are much smaller than traditional structures, weighing anything from 0.01kilograms to 180 kilograms.[6] The benefits of these smaller systems are ample, but include reduced manufacturing costs so that only modest facilities are required to construct them. As a result, a plethora of start-up companies have been attracted to their design. Furthermore, they can be mass-produced which makes it cheaper to create satellite 'constellations'—meaning a group of satellites working in concert to provide coverage for the entire Earth.[7] These constellations can provide global positioning systems, telecommunication infrastructures and remote sensing opportunities which can be used to create early warning systems for natural disasters or to anticipate food shortages. Satellite constellations have even been used to track and dismantle modern slavery operations.[8] The development of space technologies also often coincides with the development of other digital technologies which are themselves used to strengthen internal solidarity and economic growth.[9] Thus, small satellites both directly and indirectly contribute to the economic growth of a developing nation.[10]

Small satellites can also strengthen space economies outside of the context of international sustainable development. In any economic context, where competitive forces are weakened, market monopolies or oligopolies are discouraged to innovate or drive down their costs.[11] But market competition can provide a solution. For example, it was national competitive forces that induced NASA to finally open its launch programme to competitors in a move which led to technological innovation and multi-actor use.[12] Small satellite enterprise, being a comparatively inexpensive way of accessing space, can inject the competition required in an industry dominated by a handful of wealthy entrepreneurial ventures to diversify the market.[13]

Unfortunately, small satellite business occupies a precarious legal position in international space law which can impede the entrepreneurial efforts of commercial

---

[5] Sweeting M, 'Modern Small Satellites—Changing the Economics of Space' (2018) 106 Proceedings of the IEEE 3, 353.

[6] Xue Y and Yingjie L, '*Small Satellite Remote Sensing and Applications—History, Current and Future*' (2008) 29 International Journal of Remote Sensing 15, 4339.

[7] Larsen P, '*Small Satellite Legal Issues*' (2017) 82 Journal of Air Law & Commerce 75, 276; one commercial entity, OneWeb, is creating a constellation of 60 satellites to provide network coverage to the entire Earth.

[8] Boyd D, 'Analysing Slavery through Satellite Technology: How Remote Sensing Could Revolutionise Data Collection to Help End Modern Slavery' (2018) 4 Journal of Modern Slavery 2, 169–199.

[9] Schia N, 'The Cyber Frontier and Digital Pitfalls in The Global South' (2018) 39 Third World Quarterly 5, 828.

[10] For example, the Japanese Space Policy includes sections on economic and strategic growth; *see* Aoki S, '*Japanese Law and Regulations Concerning Remote Sensing* Activities' (2010) 36 Journal of Space Law 33, 350-64.

[11] Bockel J, '*The Future of the Space Industry*' (2018) 173 Economic and Security Committee 18, 4.

[12] Ward P, The Consequential Frontier: Challenging the Privatisation of Space (Melville House, 2019), 56.

[13] For example, Elon Musk's SpaceX and Jeff Bezos' Blue Origin.

space actors. I will begin my critique of space law by first setting out the broad provisions of its regime.

2.2 Public International Space Law

Upon the launch of *Sputnik 1* in 1957, space had moved to the forefront of American and Soviet efforts to exhibit technological and ideological superiority over the other. Yet, either through mutual political understanding or more probably the economic difficulty of realising large military structures in outer space, an understanding was reached between the superpowers that outer space was not to be weaponised.[14] This paved the way for political reflection on what legislative state outer space should assume and, from 1967 to 1975, five treaties were developed by the United Nations Committee for the Peaceful Uses of Outer Space (COPUOS) to create a *corpus juris spatialis*.[15] These treaties still represent the core substance of space law, alongside conventions, bilateral agreements, guidelines and national space laws which, although in their formative stage, must be considered equally important to the facilitation of private satellite enterprise. The five main treaties of international space law are as follows: The Outer Space Treaty (OST)[16]; Rescue and Return Agreement[17]; Liability Convention (LIAB)[18]; Registration Convention (REG)[19]; and the Moon Agreement.[20] Of these, the OST is the most widely ratified, and is often referred to as quasi-constitutional by commentators due to the fundamentality of its substance to outer space activities.[21]

Article I of the OST presents an egalitarian paradigm, reading: *'The exploration and use of outer space...shall be carried out for the benefit and in the interests of all countries, irrespective of their degree of economic or scientific development, and shall be the province of all mankind'*.[22] Of course, the realisation of this article is subject to the political and economic sensitivities of the dominant state space actors

---

[14] Gabrynowicz J, 'Space Law: Its Cold War Origins and Challenges in the Era of Globalization' (2004) 37 Suffolk University Law Review 1041, 1043.

[15] Malinowska K, *Space Insurance: International Legal Aspects* (Kluwer Law International, 2017), 12.

[16] Outer Space Treaty, *supra* n.3.

[17] Agreement on the Rescue of Astronauts and the Return of Objects Launched into Outer Space (Apr. 22, 1968) 19 U.S.T. 7570 [herein the Rescue and Return Agreement].

[18] Convention on International Liability for Damage Caused by Space Objects (Mar.29, 1972) 24 U.S.T. 2389 [herein the Liability Convention].

[19] Convention on Registration of Objects Launched into Outer Space (Jan. 14, 1975) 28 U.S.T. 695 [herein the Registration Convention].

[20] Agreement Governing the Activities of States on the Moon and Other Celestial Bodies (Dec. 18, 1972) 1362 U.N.T.S. 3 [herein the Moon Agreement].

[21] Robinson G and White H, Envoys of Mankind—A Declaration of First Principles for the Governance of Space Societies (Smithsonian Institution Press, 1985), 187.

[22] Outer Space Treaty, *supra* n.3, art. I.

and, therefore, Article I should be viewed as a rhetorical idealism which lacks any method for practical implementation. It is argued later that an international focus on small satellite facilitation offers one vehicle for the realisation of this article. Article II elaborates that *'[o]uter space is not subject to national appropriation by claim of sovereignty'*,[23] and thus space is granted a legal position similar to that of the high seas in that no state can extend its territorial jurisdiction into outer space.[24] Importantly, this does not defeat assertions of jurisdiction over state space objects,[25] and is restricted in the sense that activities must not frustrate general international law.[26]

It is unfortunate that the OST made little effort to legislate for commercial access to space. This was partly because the treaty was more concerned with international security at the height of the Cold War, and partly because the drafters incorrectly assumed that commercial activity in outer space was unlikely.[27] Instead, responsibility for all space activities is placed on the state,[28] and the OST calls for state 'authorisation and continuous supervision' of all national space activities irrespective of whether the activities are public or private in nature.[29] This provides an impetus for national legislators to establish domestic legal regimes which maintain tight control over private space entities, most importantly with licensing and registration obligations.

3 Antiquated Treaty Law for Small Satellite Purposes

Commercial initiatives now represent 78% of the global space economy.[30] Accordingly, investments in outer space are moving from government spending on large projects to smaller private activities.[31] This is important because it shifts the vantage point of any analysis: private operators are motivated by profit and not public or military interests. Small satellite initiatives will only be pursued should an acceptable return be made on the investment capital.[32] Therefore, to impede small satellite enterprise, space law must render returns uncertain or reduced insofar that new small

---

[23] Outer Space Treaty, *supra* n.3, art. II.

[24] Convention on the Law of the Sea (Dec. 10, 1982) 1833 U.N.T.S. 397, art 86–120.

[25] Von der Dunk F, *'Legal Aspects of Satellite Communications—A Mini Handbook'* (2015) 70 Journal of Telecommunication and Broadcasting Law 4, 1–26.

[26] Outer Space Treaty, *supra* n.3, art. III.

[27] Blout P, *'Renovating Space: The Future of International Space Law'* (2011) 40 Denver Journal of International Law & Policy 515, 516.

[28] Outer Space Treaty, *supra* n.3, art. VI.

[29] Ibid.

[30] Ward, *supra* n.12 at 61.

[31] Karacalioglu G, *'Impact of New Satellite Launch Trends on the Orbital Debris Environment'* (2016) Space Safety Magazine, https://www.spacesafetymagazine.com/space-debris/impact-new-satellite-launch-trends-orbital-debris. Last accessed 7 February 2020.

[32] Larsen P, *supra* n.7 at 281.

satellite business models are rendered economically inviable. Three key areas where space law acts as an impediment to small satellite enterprise will now be detailed, with the aim of highlighting exactly where the international legal rule translates into the commercial disincentive.

3.1 Definitional Uncertainty

Small satellites challenge some of the most fundamental elements of space law, none more so than definitional questions as to the concept of a 'launch' and the delimitation of outer space (often referred to as the 'boundary issue').[33] As an important precursor, commercial entities are not affected directly by definitional uncertainty as national activities in space are the responsibility and, potentially, liability of the state only. But the particularities of a commercial satellite, including the mode of launch and the desired orbital slot, will be important information for a state agency in deciding whether to grant an operative licence pursuant to Article VI of the OST. To trigger the applicability of the OST and LIAB, taken together, one must: "launch or procure the launch of an object into outer space".[34]

The first concept small satellite constellations challenge is the concept of a 'launch' itself. A launch, as Cheng notes, has no treaty definition and has been tested ever since the air launches of the Pegasus Vehicle in the 1980s where a spacecraft was released from underneath an airplane.[35] It was unclear whether, for the purposes of space law, rocket propulsion in some upward trajectory was required. In the case of Pegasus, the launch was at least academically considered to fulfil the criteria of a launch due to the use of basic rocket technology; however, in the case of small satellites, some commercial space actors have plans to 'release' their satellites into low-Earth orbit via third party, pre-positioned spacecraft (often called the 'two-stage-to-orbit' method). Indeed, the concept has already been used by the International Space Station where CubeSats were released from an onboard 'launcher'.[36] Furthermore, some commercial entities are planning to operate a satellite 'delivery' service whereby small satellites are brought into space via balloons.[37] At the outset, therefore, a case could be made to enshrine in public international space law a legal definition of a launch, one that caters for obscure technological developments with regards to orbit entry.

Yet, aside from the elusive concept of a 'launch', it has been correctly asserted by Christol and Del Duca that the demarcation point is still a most poignant question

---

[33] Cheng B, Studies in International Space Law (Clarendon Press, 1997), 467.
[34] Outer Space Treaty, *supra* n.3, art. VII and Liability Convention, *supra* n.18, art. 1(a).
[35] Cheng, *supra* n.33 at 467.
[36] Howell E, *'Space Station Opens Launch Pad for Tiny Satellites'* (Space.com, 2012), https://www.space.com/18098-space-station-launches-tiny-satellites.html. Last accessed 7 February 2020.
[37] Rimmington K, *'Balloon 'Taxi Service' to Take Satellites to Space'* (BBC News, 2019), https://www.bbc.co.uk/news/uk-wales-49827415. Last accessed 2 January 2021.

and an unsettled issue in air and space law.[38] The OST leaves little doubt that outer space presents a distinct legal regime[39]: Article II rejects any declaration of state sovereignty and thus *de jure* separates outer space from the airspace underneath where sovereignty reigns.[40] Previously, commentators have suggested that the delimitation issue should either be left to scientists to 'discover' at some future point,[41] or that the issue is 'not of crucial importance'.[42] Unfortunately, these arguments no longer hold weight in the context of small satellites which are traditionally placed in very low-Earth orbit (LEO). Therefore, the precise delimitation of outer space is vital to correctly trigger the provisions of the liability convention.

Two schools of thought have prevailed. The 'functionalist' school of thought—which focuses on regulating the *type* of activities irrespective of their location—has denounced attempts to create a boundary between airspace and outer space as a 'comedy of errors' owing to the unfounded theories that have been put forward by 'spatialist' commentators to devise a legally workable and scientifically accurate boundary.[43] One example, the aerodynamic-lift theory, attempts delimitation by hypothesising the moment where 'air' no longer provides support through lift for vehicles. Therefore, 'airspace ends where an aircraft will no longer find sufficient aerodynamic lift to sustain a flight'.[44] In the absence of explicit treaty definition, one may analyse any national legislation to establish the state practice and *opinio juris* necessary for the formation of customary international law. In this context, the South African Aviation Act 1962 might provide support for the aerodynamic lift theory where it defines an aircraft as a 'machine that can derive support in the atmosphere from the reactions of the air'.[45] However, modern technological revelations such as the *X-15* aircraft and *Virgin Galactic's White Knight Two* blur the distinction between 'aircraft' and 'spacecraft' with their ability to travel higher than any aircraft before them. Thus, with neither aircraft nor spacecraft given appropriate legal definitions, the aerodynamic-lift theory suffers in its conceptual plausibility.

---

[38] Christol C reviewing *'Derecho Internacional Contemporaneo: La Utilizacion del Espacio Ultraterrestreto'* (1993) 87 American Journal of International Law, 491; Del Duca P reviewing Sgrosso G, 'La Responsibilita Staki per le Attivita Svolte Nello Spazio Extra Atmosferica' (1993) 87 American Journal of International Law, 355.

[39] Von der Dunk F, *'The Delamination of Outer Space Revisited: The Role of National Space Laws in the Delimitation Issue'* (1998) 5 Space, Cyber, and Telecommunications Law Program Faculty Publications 1, 256.

[40] Outer Space Treaty, *supra* n.3, art. II.

[41] Cheng, *supra* n.33 at 121.

[42] Oduntan G, 'The Never-Ending Dispute: Legal Theories on the Spatial Demarcation Boundary Plane between Airspace and Outer Space' (1986) 1 Hertfordshire Law Journal 2, 66.

[43] McDougal M, *Law and Public Order in Space* (Yale University Press, 1963), 63.

[44] Potter P, 'International Law of Outer Space' (1958) 52 American Journal of International Law, 305; Hogan J, 'Legal Terminology for the Upper Regions of the Atmosphere and Space Beyond the Atmosphere' (1957) 51 American Journal of International Law 2, 362.

[45] South Africa Aviation Act (Act No. 74 of 1962).

The urgency for conceptually sound delimitation is stressed when one analyses the process through which customary international law is created. Indeed, the International Court of Justice (ICJ) in the *North Sea Continental Shelf Case* contended there might be circumstances where states that were 'specially affected' by a purported rule should be given special consideration when determining the content and existence of customary international law.[46] To assimilate this idea into the context of space exploration, the interests of states who already possess the technological capacity to undertake space activities would be prioritised to the detriment of prospective space actors. On one hand, a degree of immediate conformity may be a more attractive prospect compared to new space actors ignoring the customary principles of space activities developed by established space actors, as happened with the illustrative example of the developing nations keen to explore Antarctica who accepted the role of 'Consultative Parties' to the 1959 Antarctic Treaty[47] rather than devising their own take on its legal status.[48] However, although this point was evaded by the Court in the *Nuclear Advisory Opinion*,[49] the realisation of 'special consideration' rules in the space context perpetuates a regime of dominance and monopoly, allowing major space powers a 'respectable disguise' to preserve their vested interests in the name of special affectedness.[50] If the diversification of space actors is a genuine concern of the international community, the interests of developing space nations should also be considered as relevant to the formation of customary international law.

Further troubling is the prospect of 'instant customary law', a development in international jurisprudence which suggests that nascent areas of practice may develop customary rules quicker than established areas.[51] As Cheng contends, this would apply to the nascent realm of outer space and particularly any commercial operations undertaken there.[52] One small satellite, without the capabilities of manoeuvre, may be seized or destroyed for trespass in another's jurisdiction simply because a major space power declares the height of its national airspace, as has happened before with maritime boundaries.[53] This would obviously be an onerous rule for small satellite operators who possess constellations spanning above multiple jurisdictions. Fortunately for small satellite purposes, 'instant customary law' is a controversial rather than orthodox notion.[54] Nonetheless, the uncertainty fostered by the open-ended questions of special affectedness and instant custom has a disincentivising effect on business. Indeed, any uncertainty will inevitably increase the rate of business inaction, where investment is zero, given that private entities prefer to 'wait and see'

---

[46] North Sea Continental Shelf Judgement, ICJ Reports 1969, 3, para [74].

[47] Antarctic Treaty (1. December 1959) 402 UNTS 71.

[48] Von der Dunk F, *'Advanced Introduction to Space Law'* (Edward Elgar 2020), 8.

[49] *Legality of the Threat of Nuclear Weapons*, Advisory Opinion, ICJ Reports 1996, para [226].

[50] Danilenko GM, Law Making in The International Community (Nijhoff, 1993), 96.

[51] Cheng, *supra* n.33 at 647.

[52] Ibid., 648.

[53] Ibid.

[54] Crawford J and Viles T, 'International Law on a Given Day' in Ginther K, Völkerrecht Zwischen Normativen Anspruch und Politischer Realität (Duncker & Humblot, Berlin 1994), 93.

rather than undertake costly action with uncertain consequences.[55] Therefore, the lack of definitional clarity for the central components of a 'launch' and the demarcation issue are the first elements of international space law to act as an impediment to prospective small satellite enterprise.

3.2 Satellite Insurance

Historically, insurance has been integral to any private innovation, growth and investment in space enterprise as it facilitates increased financial certainty,[56] a key factor in the realisation of space activities for profit-motivated entities. However, insurance coverage for many commentators is also the main limiting factor to the growth of the small satellite industry.[57]

As detailed above, Article VI of the OST obliges states to 'authorise' and 'continuously supervise' their national space activities. This authorisation process protects governments from incurring liability via the mandatory insurance coverage which indemnifies a government for any damage caused on Earth or in outer space.[58] In several states, this obligation to purchase insurance is not explicitly implemented by national space law but constitutes an administrative condition to obtain a licence to conduct space activities.[59]

This legal regime means that smaller prospective space actors are often required to cover onerous insurance charges to obtain a license. Moreover, high insurance costs are often disproportionately skewed to the detriment of smaller companies as larger companies have the large capital reserves necessary to self-insure. Improvements have been made, in the UK for example, with the introduction of an 'indemnity cap' at €60 million via the Deregulation Act 2015.[60] However, a license only applies to one satellite and hence an operator attempting to launch a small satellite constellation will face an insurance fee disproportionate to the scale of their activity. Once a facilitator of space innovation, insurance now weakens the financial efficacy of small satellite business modelling.[61]

It should be stressed that international space law does not *directly* impede small satellite enterprise. Indeed, the outer space treaties are silent on insurance coverage.[62]

---

[55]Bloom N, Bond S and Reenen JV, *'Uncertainty and Investment Dynamics'* (2007) 74 Review of Economic Studies 2, 391–415.

[56]Chaney L and Hirano N, *'End of an Era? Satellite Insurance Faces Changing Landscape'* (2019) 5 Centre for Space Policy and Strategy 2, 7.

[57]Hameed H, 'Small Satellites—Entrepreneurial Paradise and Legal Nightmare' (2018) 8 Journal of Space Technology 1, 74.

[58]Larsen P, *supra* n.7 at 292.

[59]Malinowska, *supra* n.15 at 34; this is the situation in states like Canada, Norway, South Africa and the UK.

[60]Deregulation Act 2015, s12(5) amending the Outer Space Act 1986, s10.

[61]Science and Technology Committee, Satellites and Space (HC 2016–17, 160), para [78].

[62]Malinowska, *supra* n.15, 45.

However, it will be obvious that those states who establish national licencing regimes usually and explicitly also include mechanisms for the reimbursement by licensees of any compensation paid. Insurance requirements are therefore an *indirect* result of a state-centric liability regime which has submitted non-state actors to the interests of their respective state through licensing procedures.

Space debris, discussed next, complicates this situation as a greater number of space objects renders collision more likely. Larsen explains that governments who seek to shift this increased liability exposure onto private insurance companies will not succeed given the limited pool of insurance available for outer space activities.[63] Thus, governments will have to restrict the licences it offers to prospective small satellite businesses to limit their liability. This is likely to have a disproportionate effect on unknown, start-up space actors over the experienced oligopolies already in space and hence any national diversification of space actors is rendered increasingly unlikely. Correctional methods to the debris problem will be discussed next.

3.3 Space Debris Remediation

Low-Earth Orbit (LEO) is currently an outer space graveyard. It is estimated that there are approximately 23,000 pieces of 'space junk' in LEO measuring 10cm or more, with millions of smaller pieces in tow.[64] These pieces of debris include pieces of rocket, satellite and other clutter, circling the Earth at almost seven times the speed of a bullet, ready to destroy or displace artificial satellites or inflict fatal injuries to astronauts.[65] The problem is further complicated by the Kessler Syndrome, a theory which contends that space debris will continue to create more debris as it cumulatively collides, fragmenting endlessly into smaller and more pieces and eventually impeding access to outer space altogether.[66]

Small satellites obviously multiply the problem. There are plans to triple the number of operating small satellites in LEO,[67] and each addition adds more variables to be taken into account when maintaining order in outer space.[68] The influx of

---

[63] Larsen, *supra* n.7 at 293.

[64] Kumar A, *'The Trouble with Space Junk'* (The Economist, 2015), https://www.economist.com/the-economist-explains/2015/05/10/the-trouble-with-space-junk. Accessed 11 March 2020.

[65] Ibid.

[66] Johnson N, *'The Historical Effectiveness of Space Debris Mitigation Measures'* (2005) 11 International Space Review 6, 9. Recent incidents include the 2007 destruction of the Fengyun-1C weather satellite during an anti-satellite missile test by China, which created a massive amount of orbital debris (approximately 36% of all orbital debris currently in low Earth orbit). Similarly, in 2019 India announced the successful completion of an anti-satellite missile test, creating a new 400-piece mass of debris which increased the risk of impacts to the International Space Station by 44% over 10 days.

[67] Karacalioglu, *supra* n.31.

[68] Larsen, *supra* n.7, 294.

satellites from oligopoly space actors will inevitably reduce the consolidated opportunities of all prospective small satellite enterprises to use space. Some measures have been introduced by way of 'soft law' to *mitigate* further debris being created, although, as Liou and Johnson correctly submit, the only way to prevent future problems for the commercialisation of space is to remove the existing space debris through remediation measures.[69] Unfortunately, international space law cannot, at least in its current form, facilitate space debris remediation (SDR) programmes to remove debris from LEO.

Liability and responsibility are central concerns when analysing the effects of environmental harm in space. The general rule is that states bear liability when deemed the 'launching state', where its '[space] object or component parts' cause damage.[70] The LIAB creates an absolute liability regime for damage on Earth or to aircraft and a fault-based liability regime for damage caused in outer space.[71] This legal paradigm creates two major uncertainties, namely the concepts of 'space debris' and 'fault'. First, there is no internationally binding definition of 'space debris' in the UN treaties. This is because, at the time of their conclusion, environmental concerns were not at the forefront of state agendas.[72] Therefore, on one reading of the space treaties, all debris (even paint flecks) are maintained as 'space objects' under the jurisdiction of the state that launched them, and this diminishes the feasibility of projects[73] designed to capture or re-orbit debris as the unilateral removal of a foreign space object is considered a breach of sovereignty and a possible act of war.[74]

Commentators have put forward solutions to this 'continuing ownership' problem and hence incentivise SDR action. Muñoz-Patchen, for example, has applied the Anglo-American common law doctrine of the right to abandonment to space debris.[75] In her hypothesis, space debris is a worthless material similar to litter which can be abandoned by any 'voluntary act' of the owning state.[76] Any identifiable debris should be considered abandoned if the UN register, which states must submit their space activity details to pursuant of the Registration Convention,[77] represented that the debris was no longer functional. This solution could avoid the fear of geopolitical disputes over the ownership of space debris and incentivise state or commercial clean-up operations.

---

[69]Liou J and Johnson N, *'Risks in Space from Orbiting Debris'* (ScienceMag, 2006), https://science.sciencemag.org/content/311/5759/340. Last accessed 10th March 2020.

[70]Outer Space Treaty, *supra* n.3, art VII.

[71]See Liability Convention, *supra* n.18, art III.

[72]Larsen, *supra* n.7 at 294.

[73]Harpoons, Space Lasers and Nets have all been suggested; *see* Tallis J, *'Remediating Space Debris: Legal and Technical Barriers'* (2015) 9 Strategic Studies Quarterly 1, 92.

[74]Cheng B, supra n.33 at 647; also see Weeden B, 'Overview of The Legal and Policy Challenges of Orbital Debris Removal' (2011) 27 Space Policy 1, 41.

[75]Muñoz-Patchen C, 'Regulating the Space Commons: Treating Space Debris as Abandoned Property in Violation of the Outer Space Treaty' (2018) 19 Chicago Journal of International Law 1, 233–259.

[76]Ibid., 250.

[77]Registration Convention, *supra* n.19, arts. II(I) and IV.

However, abandonment suffers from both legal and political problems. Firstly, there is no international consensus on the right to abandonment. Even in Anglo-American Jurisprudence, the right is often overlooked and assimilated with a serve to determine ownership.[78] Hence it is unlikely that the right to abandonment is a norm with international force even if the right can be found in a handful of common and civil law jurisdictions.[79] Secondly, the RC does not mandate states to update the register to include information about the operational capacity of their satellite. The 'voluntary act' necessitated by the doctrine of abandonment instead relies on the political will of states to permit the capture, examination and use of their space technologies; an unlikely event given the strategic and military sensitivity of satellite assets.[80] Moreover, the technology that SDR requires may furthermore be used as anti-satellite weaponry, a factor which only exacerbates the political reluctance of states to incentivise SDR operations.[81]

This writer submits that self-help in a state of necessity might fare better when invoked by a state entity to justify its national commercial SDR measures to clean-up the space environment.[82] The International Court of Justice in the *Gabčikovo-Nagymaros Project* Case observed that necessity as a ground for precluding wrongfulness can be accepted under strictly defined conditions within international environmental law.[83] Such conditions, adapted for our purposes, would be the imminent peril of the space environment in order to preserve its usability. Debris poses an imminent threat to the interests of all states yet, even if collisions were only likely upon the insurgence of small satellites into outer space, it is not fatal that the 'peril' is far off as the danger need only be 'inevitable'.[84] Unfortunately, the doctrine of necessity is dependent on a commercial, non-commercial actor or both being the first to undertake an SDR operation and is thus entirely dependent on an actor's confidence in the doctrine.

Moreover, fault-based liability for space actors is problematic. More often than not, a single piece of debris cannot be identified to apportion liability to the state that launched the space 'object'. Even if the debris can be identified, there exist no space traffic rules which set standards for conduct in space—a breach of which

---

[78] Strahilevitz L, *'The Right to Abandon'* (2010) 158 University of Pennsylvania Law Review 355, 363–64; Peñalver E, *'The Illusory Right to Abandon'* (2010) 109 University of Michigan Law Review 191, 204.

[79] *See Armory v Delamirie* [1722] EWHC KB J94; *Stewart v. Gustafson* [1998] 171 Sask. R. 27 (CA); Akkermans B, *Cases, Materials and Text on Property Law 1010* (Hart Publishing, 2012), 546.

[80] Popova R, 'The Legal Framework for Space Debris Remediation as a Tool for Sustainability in Outer Space' (2018) 5 Aerospace 2, 10.

[81] Ibid.

[82] Articles on State Responsibility for Internationally Wrongful Acts (Dec. 12th, 2001) UNGA Res. 56/83, art 25.

[83] Gabčikovo-Nagymaros Project (Hungary v Slovakia), Judgment, I.C.J. Reports 1997, [7].

[84] Ibid., para [54].

could indisputably be considered as establishing fault under the Liability Convention.[85] Unfortunately, no standard of care is enshrined, and the issue has not been brought before an international tribunal. Hence, all accusations of 'fault' are merely ideological assertions of what should encompass a failure of a state's duty of care. Although, as Jakuh contends, there is a general notion of 'fault' in international law,[86] and Article III of the OST is generally interpreted as a 'fall-back' clause which may substitute in general international law when space law does not provide an answer to a legal question,[87] there is still warranted doubt over its applicability.

Tallis correctly remarks that the remediation of space debris meets a major obstacle in a 'perplexing legal regime that makes incentivising through liability and ownership laws ambiguous and difficult to enforce'.[88] In this context, it is clear that the international community should prioritise a binding separation between a 'space object' and 'space debris', and a new legal code developed around the rights and duties of 'launching states' for the debris they create. For small satellites, the lack of SDR incentivisation not only decreases the likelihood of *new* small satellite enterprise but also increases the commensurate risk to small satellite activities. This writer agrees with the European Space Agency's chief debris expert that it would be 'insane' to allow an increase in the present levels of space debris by adding large numbers of small satellites into outer space.[89]

4 The Way Forward

It has been depicted how international space law has a disincentivising effect on small satellite enterprise. If we aim to diversify space actors and reduce oligopolistic control by stimulating and strengthening international space economies, then space law must be improved to better reflect the space age of the twenty-first Century. Next, correctional means to achieve this aim will be explored.

---

[85] Von der Dunk F and Tronchetti F, *Handbook of Space Law* (2nd edn, E Elgar 2017), 278.

[86] Jakuh R, 'Iridium-Cosmos Collision and its Implications for Space Operations' in Kai-Uwe Schrogl, Yearbook on Space Policy: 2008/2009 (Springer, 2010), 225.

[87] Von der Dunk F, *'Advanced Introduction to Space Law'* (Edward Elgar 2020), 19.

[88] Tallis, *supra* n.73 at 91.

[89] H Krag, 'International Aeronautical Congress Panel Discussion on the Projection and Stability of the Orbital Debris Environment in Light of Planned Mega-Constellation Deployments' (2016), https://www.iafastro.org/wp-content/uploads/2014/04/The-Orbital-Debris-Environment-In-The-Light-Of-Planned-Mega-Constellation-Deployments-at-IAC-2016.pdf. Last accessed 10th February 2020.

4.1 Legal Personality

This paper has suggested that insurance requirements, indirectly imposed by a state-centric liability regime, presented a financial obstacle to small satellite enterprise. However, alternatives are unlikely. One solution might be a liability regime which instead grants international legal personality to space corporations, as has been manifesting in the context of human rights law.[90] This model might lessen the financial burden of insurance costs on small satellite operators as they could bypass state agencies who must, as a matter of public policy, guard against any and all foreseeable risks. However, international legal personality should be dissuaded in our case. For one, even passive *ratione personae* for multinational corporations in the context of human rights is a controversial notion and, as Trechsel submits, is currently unrealistic.[91] It is therefore unlikely to be extended to space law and small entrepreneurial start-ups. Moreover, taking the supervisory onus away from states would require treaty amendments more fundamental than should be politically expected at this stage, and risk opening up the fragile tenants of international space law only to destroy its customary foundations and weaken confidence in its authority; as Von der Dunk submits, parties risk losing more than they would gain.[92] This writer submits that, for the problem of onerous insurance burdens, lobbying political institutions to include waivers or favorable conditions in their licensing requirements for small satellite activities would be more effective, as is the case in Austria where small satellite missions can be fully exempted from the insurance requirement should it be in the 'public interest'.[93]

4.2 Soft Law Mechanisms

It was explained why SDR efforts are essential for any prospective space actor diversification through small satellite enterprise. But what solutions can international law offer, if any? We shall begin with 'soft' provisions before evaluating 'hard' provisions, concluding that, given the seriousness of the debris issue, binding law should be advanced by the COPUOS. Abbot and Snidal's definition of 'soft law' as

---

[90] J Kyriakakis, 'International Legal Personality, Collective Entities and International Crimes' in Gal-Or N, Ryngaert C and Noortmann M, Responsibilities of the Non-State Actor in Armed Conflict and the Market Place: Theoretical and Empirical Findings (Brill, 2015), 83.

[91] Trechsel S, *'A World Court for Human Rights?'* (2004) 1 North western Journal of International Human Rights 1, 11.

[92] Von der Dunk F, *'Advanced Introduction to Space Law'* (Edward Elgar 2020), 25.

[93] Austrian Federal Law on the Authorisation of Space Activities and the Establishment of a National Space Registry (Austrian Outer Space Act, adopted by the National Council on 6 December 2011, entered into force on 28 December 2011), s (4).

meaning legal arrangements which are weakened along the dimensions of 'obligation, precision, and delegation' shall be used.[94]

The Space Debris Mitigation Guidelines (SDMGs) are the leading international arrangement to deal with space debris.[95] The guidelines are concerned with the mitigation of debris by prospective space actors (or existing ones 'if possible') and convey seven steps that an actor can take to minimise the harmful by-products of their space activity including de-orbiting rules[96] and component specifications.[97] These guidelines include a definition of space debris as 'non-functional' objects, a helpful clarification which goes some way to rectifying problems of definitional uncertainty.[98] The guidelines are not concerned with the remediation of the space debris as discussed above, however, they might be used as an example of how 'soft' regulation could invigorate international consensus around the need for SDR agreement and prepare the ground towards the creation of customary international law.

Unfortunately, the specificity of the problem of space debris does not lend itself towards the formation of customary international law. The requirements for rules to attain customary status are some repeated practice over time which is extensive, virtually uniform and perceived as obligatory,[99] and although there have been some marked improvements in the determination of space faring states to use the Guidelines in their domestic authorisation processes pursuant to the liability convention, as Imburgia comments, the prevailing practice of the modern space age amongst space faring nations is to limit the creation of new orbital debris when it is cost-effective to do so, and not to limit oneself to the sustainable exploitation of space benefits.[100] Therefore, although mitigation standards have improved on average, the practice is by no means extensive[101] and is not conducive to the long-term sustainability of outer space activities or the objective of ensuring that Earth's orbits are safe for commercial use. It is submitted that, if soft, non-binding measures were to be introduced to tackle the problem of space debris remediation, they would be subject to similar criticisms in the sense that the extensive practice of states might orient towards a commercial, for-profit solution which would either exacerbate or delay the problem rather than present a sustainable solution. Moreover, in the absence of 'instant customary law', the time non-binding rules take to translate into a customary

---

[94] Abbott K and Snidal D, '*Hard and Soft Law in International Governance*' (2000) 54 International Organizations 3, 421–422.

[95] Space Debris Mitigation Guidelines of the Committee on the Peaceful Uses of Outer Space, UN OOSA (2010), ST/SPACE/49, https://www.unoosa.org/pdf/publications/st_space_49E.pdf. Last accessed 11 February 2020.

[96] Ibid., Guideline 6.

[97] Ibid., Guideline 1.

[98] Ibid.

[99] *See North Sea Continental Shelf*, Judgement, I.C.J Reports 1969, 3.

[100] Imburgia J, 'Space Debris and Its Threat to National Security: A Proposal for a Binding International Agreement to Clean up the Junk' (2011) 44 Vanderbilt Journal of Transnational Law 589, 624.

[101] Von der Dunk and Tronchetti, *supra* n.85 at 757.

international agreement would require time that Kessler's hypothesis suggests is not there, especially if instances like India's ASAT-test are to become more frequent.[102]

4.3 Hard Law Mechanisms

Binding resolutions are necessary where political sensitivities merit State inaction on the issue of space debris remediation. What is needed is a binding treaty obligation on states which rectifies the problem of continuing ownership, financially supports SDR and defines the terms 'outer space' and 'space debris'. It is understood that the conclusion of a new international agreement would be politically difficult, and that an agreement which caters to the discretion of states to implement their own solution to the problems of public international space law is preferable to some commentators. Nonetheless, it is hoped that the seriousness of the space debris issue will incentivise political bodies to cooperate. The only articles extraneous to this cause would be definitional clarifications to support small satellite entrepreneurship.

The Legal Subcommittee (LSC) of UNCOPUOS are in the pertinent position to propose this treaty. Substantively, the committee should declare that international space law as it stands is insufficient to deal with the problem of space debris. It should then adopt a position that first prioritises facilitative SDR measures, regardless of the current technological and economic feasibility of such operations. This might include measures to 'abandon' property through the Registration Convention as discussed above or provide reassurance so that the doctrine of necessity precluding wrongfulness is rendered unnecessary.

Second, the LSC should demand the imposition upon all space-faring nations the responsibility, upon ratification, to contribute money to an international fund. It is recognised that SDR operations might not be able to support themselves financially in the short-term. Therefore, a clean-up fund similar to that of a national recycling scheme might be called for.[103] Options include a 'space access fee', as proposed by Ackers, which would likely manifest as a condition for license from national space agencies. The benefits of the 'access fee' would be that it only imposes liability on actors who currently use space. However, it discriminates against those states who have not previously engaged in space activities and, for our purposes, actually imposes more of a financial burden on small satellite start-ups. This writer instead advocates for Imburgia's 'market-share liability' model which bases a state's contribution amount on their responsibility for the debris currently in orbit[104]; a fair

---

[102] Tellis A, *'India's ASAT Test: An Incomplete Success'* (Carnegie Endowment for International Peace, 2019), https://carnegieendowment.org/2019/04/15/india-s-asat-test-incomplete-success-pub-78884. Last accessed 4th January 2021.

[103] See Akers A, 'To Infinity and Beyond: Orbital Space Debris and How to Clean It Up' (2012) 33 University of La Verne Law Review 285, 312; also see Tian Z and Cui Y, 'Legal Aspects of Space Recycling' in Froehlich A, 'On-Orbit Servicing: Next Generation of Space Activities' (Springer, 2020), 35.

[104] Imburgia, *supra* n.100 at 629.

and effective resolution to the debris problem. Although this resolution would be economically burdensome on the United States, the removal of space debris is in the long-term national security interest of the United States.[105] Market-share liability would also work towards the aim of diversification, as emerging space actors would only have to pay for the debris they subsequently create. The LSC should finally advocate for definitions of 'launch', 'space' and 'space debris' to be included in the final treaty, clearing up any definitional uncertainty created by the OST and LIAB.

This article does not seek to include a discussion of the complete legal ramifications of this 'hard' law mechanism. Nonetheless, it is apparent that binding international obligations are the only effective resolution to this problem and might even be considered an 'unavoidable' next step for international agreement concerning debris remediation.[106] Soft law has been the tool of choice for modern space regulation; indeed, the recent Long-Term Sustainability Guidelines are demonstrative of a renewed commitment by UNCOPOUS to the tools of soft law for generating awareness, exemplary practice and international commitment to the ideals of space sustainability. However, it is submitted that states need to be even more ambitious in reacting to this threat to humanities' continued presence in outer space.

5 Conclusion

The legislators of the 1960s referred to outer space as 'the province of all mankind' preserved for the 'interests of all countries'.[107] Unfortunately, notwithstanding this rhetoric, a 'space gap' has been preserved which apportions the benefits of space technologies unequally between nations. This situation is perpetuated by public international space law which, as this article has shown, impedes access to space for prospective small satellite users hoping to stimulate and strengthen their respective space economies. This argument easily leads to a demand for immediate and unrestricted use of this technology for its egalitarian benefits. However, small satellite technology has the potential to exacerbate the space debris problem and, within a legal regime that blocks remediation efforts, it would not be in the interests of small satellite entrepreneurs and the states that control their activities to ignore this danger. Solutions can only originate in the global forum and it has been submitted that the most efficient correctional mechanism would be a supplementary treaty which reaches a multilateral consensus on a set of standards to eradicate uncertainty, the fear of legal reprisal against those who seek to fix the problem and economically incentivise remediation efforts. This is recognised to be the most politically laborious option; however, the status quo will not survive when defence communities realise

---

[105]Ibid., 630.

[106]Degrange V, *'Active Debris Removal: A Joint Task and Obligation to Cooperate for the Benefit of Mankind'* in Froehlich A, Space Security and Legal Aspects of Active Debris Removal (Springer, 2019), 8.

[107]Outer Space Treaty, *supra* n.3, art.1.

the unavoidable threat to their national security. Thus, states should welcome the opportunity to protect their interests *alongside* allowing for international economic liberation and space actor diversification.

Michael Gould is a LL.B. graduate from the University of Bristol Law School. With a keen interest in space law, policy and business, Michael has completed professional experience in this field with the Satellite Applications Catapult in the United Kingdom and is now working as a paralegal with First Steps Legal to provide accessible support to start-ups in the space industry.

Approaches to and Loci for Regulation of Large and Mega Satellite Constellations

Jack B. P. Davies and Jonathan Woodburn

Abstract The announcement, planning, and initial deployment of third generation large and mega SC by several private commercial actors both raises novel concerns regarding the long-term sustainability of human space affairs and brings existing issues into sharper relief. Whereas there has been increasing focus on these challenges and their potential regulatory solutions within academia, governmental and inter-governmental spheres, the resultant efforts have so far been somewhat uncoordinated and ad hoc, without a guiding strategy incorporating effective regulatory design and regulatory locus selection. In this article, the authors identify and apply an underlying conceptual framework of regulatory design, using this framework to examine the relative suitability of various potential loci for regulation within the industry, national and international levels. The authors begin by proposing and examining their conceptual framework, building on previous interdisciplinary literature within the field of regulatory design. This framework is then applied within a discussion of the various bodies and institutions capable of acting as loci for regulation within the space sector. The authors find that various institutions, due to their given characteristics, are better or worse suited to facilitating various regulatory approaches. These findings are then applied to three case study issues concerning large and mega SC: radio frequency spectrum congestion; interference with optical and radio astronomy; and the production of space debris. The authors find that application of the conceptual framework and thorough analysis of various potential loci for regulation can facilitate efforts to pursue and secure regulatory solutions to these challenges. In particular, this analysis contributes real-world value in identifying the relative trade-offs inherent in the selection of different regulatory approaches, and in highlighting that a regulatory strategy involving multiple concurrent and coordinated efforts to secure multiple regulatory instruments is likely to be the most effective strategy.

J. B. P. Davies (✉)
Centre for Global Challenges, Utrecht University, Utrecht, Netherlands

J. Woodburn
European Space Agency, European Space Research and Technology Centre, Noordwijk, Netherlands

1 Introduction

Beginning in the late 1990s, a number of private companies announced plans to manufacture, launch and operate satellite constellations (SC),[1] groups of satellites of a "similar type and function, designed to be in similar, complementary, orbits for a shared purpose, under shared control".[2] These proposals fundamentally differed from previous, first generation[3] SC, such as the Global Positioning System (GPS), Galileo and GLONASS in three key aspects. First, they were intended to provide communications services rather than geopositioning. Second, they would be deployed into low Earth orbit (LEO) rather than geosynchronous Earth orbit (GEO). Finally, it was non-state commercial actors (private companies), rather than states, which planned to build and deploy them.

The planned function of this second generation of SC, namely low-latency satellite communications, meant that, in order to provide a consistent service, their scale had to greatly exceed that of the GEO-based geopositioning SC. Whereas in GEO a constellation of just three satellites is sufficient to ensure global coverage, a SC deployed in LEO requires a far greater number to achieve the same effect.[4] Put simply, being positioned closer to the Earth's surface means that satellites in LEO have a reduced surface coverage area compared with those in GEO. This challenge is overcome by greatly increasing the number of orbiting satellites in the SC, adding more 'spotlights' until universal coverage is achieved.

The need to deploy communications SC into LEO is driven by the services they aim to provide. Whereas GEO orbits offer a communications Round Trip Time (RTT) of around 600 ms, LEO orbits can offer a much more attractive RTT of 40 ms.[5] GEO latency may be sufficient for some communications activities, but it is simply too high for activities requiring a low latency, such as online video calling, and here LEO-based SC provide a competitive commercial edge. As Telesat have put it, "You can buy your way out of bandwidth problems; but latency is divine".[6]

---

[1] Notable examples include: Iridium, Globalstar and Teledesic, however by 1996 there were at minimum 16 proposed constellations accounting for 1315 satellites (See Capella, M., 'The Principle of Equitable Access in the Age of Mega-Constellations', in Froehlich, A. (ed.) *Legal Aspects Around Satellite Constellations*', Springer, 2019, p. 17).

[2] Wood, L., *'Satellite constellation networks'*, Internetworking and Computing over Satellite Networks, Springer, Boston, MA, 2003, pp. 13–34.

[3] At times, the authors refer to 'first generation', 'second generation' and 'third generation' SC, as an informal taxonomy used to differentiate between distinct historical phases in SC development.

[4] See e.g. Wood, L., 'Introduction to satellite constellations orbital types, uses and related facts', 2006, https://savi.sourceforge.io/about/lloyd-wood-isu-summer-06-constellations-talk.pdf (all websites cited in this publication were last accessed and verified on 16 January 2021).

[5] OmniAccess, 'Low Earth Orbit satellites, improving latency', www.omniaccess.com/leo/#:~:text=The%20GEO%20latency%20is%20of,and%20an%20essential%20part%20if.

[6] Telesat, 'Real-Time Latency: Rethinking Remote Networks', www.telesat.com/wp-content/uploads/2020/07/Real-Time-Latency-Rethinking-Remote-Networks.pdf.

Teledesic,[7] founded in the 1990s to build a commercial broadband SC, initially published plans for an ambitious 840-unit constellation capable of providing high upload and download speeds. The proposal was later revised in 1997 to 288 units, still at that time the biggest ever planned. Two other notable initiatives, Iridium[8] and Globalstar,[9] planned for constellations of 66 and 48 satellites respectively, still exceeding GPS (24 unit), Galileo (30) and GLONASS (24). Despite technical success in launching and successfully bringing their SC online, both Iridium and Globalstar failed to succeed commercially, finding little market demand for their services, and both companies later filed for bankruptcy. Following these failures, Teledesic further revised its plans down to just 30 units placed in medium Earth orbit (MEO), before formally ending work on the SC in 2002.[10] Both Iridium and Globalstar retained operation of their already deployed SC, going on to emerge from bankruptcy, and in recent years replacing their original satellites with next generation units.[11] Of the second-generation SC none exceeding the 100-unit threshold was ever completed.

Between 2015–2020, following a decade of disinterest in commercial LEO communications SC, a new cohort of private companies once again announced plans for ambitious SC.[12] These third generation SC are similar in function and orbit to their predecessors, however, whereas Iridium and Globalstar never exceeded 100 total satellites in orbit, these SC are planned to number well into the 100s or 1.000s of units.

Most notable of these is Starlink.[13] Starlink was first announced by SpaceX in 2015, with an initial plan for a 4.425-unit SC later supplemented by a second, 7.518-unit request. In 2018 SpaceX was granted approval for both requests, resulting in

---

[7] See https://personal.ee.surrey.ac.uk/Personal/L.Wood/constellations/teledesic.html for further information on Teledesic.

[8] See https://personal.ee.surrey.ac.uk/Personal/L.Wood/constellations/iridium.html for further information on Iridium.

[9] See https://personal.ee.surrey.ac.uk/Personal/L.Wood/constellations/globalstar.html for further information on Globalstar.

[10] Goodwins, R., 'Teledesic backs away from satellite push', 2002, www.zdnet.com/article/teledesic-backs-away-from-satellite-push/.

[11] Globalstar finished launching its second generation satellites in February 2013 (see https://spaceflightnow.com/soyuz/st26/130206launch/); Iridium finished replacing its original satellites with next generation models in January 2019 (see https://investor.iridium.com/2019-01-11-Iridium-Completes-Historic-Satellite-Launch-Campaign).

[12] See www.newspace.im/, a comprehensive database created and maintained by Kepler Communications Systems Engineer Erik Kulu. It should be noted that although the vast majority of third generation SC are being proposed and developed by private companies, there are some notable state led efforts, including the Chinese GW-A59 and GW-2 (see https://spacewatch.global/2020/10/china-pushes-ahead-with-giant-broadband-satellite-constellation/), and the Russian 'Sphere' (see https://arstechnica.com/science/2021/01/russia-may-fine-citizens-who-use-spacexs-starlink-internet-service/).

[13] See www.starlink.com/ for further information on Starlink.

an SC of over 12.000 satellites spread across multiple LEO orbital shells.[14] SpaceX have since requested approval to expand the constellation up to 41.493 units.[15] The scale of Starlink represents a major leap in LEO SC design, so much so that the authors feel it necessary to consider the terminology used to describe such systems.

The term 'constellation' is not precise enough to discern between SC of a few dozen individual satellites (such as GPS and Galileo) and ones that include 100s or 1.000s. This problem is recognized both in the space industry and academia, though without a definitive solution. The International Telecommunications Union (ITU) refers to Starlink-scale SC as 'mega-constellations'[16], while the Federal Communications Commission (FCC), National Aeronautics and Space Administration (NASA) and others use 'large-constellation'.[17] For the purposes of this article, the authors hereafter adopt a taxonomy of small SC (1–99 units), large SC (100–999 units) and mega SC (1.000 units and up). For this analysis, the authors focus exclusively on the latter two categories, large and mega SC.

Of a total of 32 announced large and mega SC (25 large and seven mega SC), one is fully launched and being replenished, seven are in the process of being launched, twelve have launched prototype satellites, eight are currently developing prototype satellites, and three have an unknown status.[18] At the time of writing, the current number of satellites launched as part of these LEO large and mega SC is 1.745/69.098 (787/6.682 units for large SC, 958/62.416 for mega SC). Although there are not yet any mega SCs fully deployed in orbit as of January 2021, this will soon no longer be true as companies continue to launch and successfully place in orbit operational satellites. SpaceX has thus far launched 953 of 41.493 planned units, with a further six launches of 60 satellites each planned through the end of February 2021.[19]

It is far from certain that all these planned constellations will be launched in their entirety, if at all. Granted, it is less likely that current efforts will fail due to the same issues as Iridium and Globalstar, as the market demand for global high-speed internet has vastly increased since the early 2000s. Investment bank Morgan Stanley estimate that by 2040 the market for satellite internet will exceed 400 billion dollars[20], a staggering figure that, if realized, will surely provide ample opportunity for numerous large and mega SC. Additionally, a number of deeper transformations within the space industry have made the manufacture and deployment of large and

---

[14]Grush, L., 'FCC approves SpaceX's plan to launch more than 7000 internet-beaming satellites', *The Verge*, 2018, www.theverge.com/2018/11/15/18096943/spacex-fcc-starlink-satellites-approval-constellation-internet-from-space.

[15]Henry, C., 'SpaceX submits paperwork for 30000 more Starlink satellites', *SpaceNews*, 2019, https://spacenews.com/spacex-submits-paperwork-for-30000-more-starlink-satellites/.

[16]See e.g., www.itu.int/en/mediacentre/Pages/2019-PR23.aspx.

[17]See e.g., https://ntrs.nasa.gov/citations/20190025975.

[18]See www.newspace.im/.

[19]Clark, S., 'Launch Schedule', *Spaceflight Now*, https://spaceflightnow.com/launch-schedule/.

[20]Warner, K., 'Low Earth orbit: Why the next big innovation battleground is out of this world', *The National News*, 2020, www.thenationalnews.com/business/technology/low-earth-orbit-why-the-next-big-innovation-battleground-is-out-of-this-world-1.1067866#:~:text=Morgan%20Stanley%20estimated%20that%20the,global%20space%20industry%20that%20year.

mega SC more feasible than ever before. Specifically, the miniaturization of space technology and advances in launch capabilities have each played a major role.

As a result of improvements in component technology, satellite design has seen a significant decrease in size without degradation in performance. One notable example is the CubeSat, a standard for miniaturised spacecraft used primarily for scientific purposes which can be made up of one or more 10 × 10 × 10 cm 1.33 kg cubic units.[21] The CubeSat's impressively small form factor, use of commercial-off-the-shelf (COTS) components, and their ability to piggyback on other missions as secondary payloads means that they are incredibly cost effective in design, manufacturing, launching and orbit insertion. As a result, these satellites offer actors on a tighter budget a realistic route to achieving a foothold in space.

Another benefit of compact satellites is the ability to deploy multiple units from a single launch vehicle. That, on top with technological advancements such as booster landing and re-use, has drastically reduced the price of launches and the cost-per-weight ratio of payloads. SpaceX is at the forefront of these advancements, with Falcon Heavy LEO launches averaging a payload cost of $1.500/kg, compared to Ariane 5G's $10.200/kg.[22] The result of this decrease in the cost of access to space is a proliferation in the number of private companies able to launch commercial activities in LEO. Nonetheless, some of the current large and mega SC ventures can and will fail.

OneWeb, one of the better-known SC, is an apt example of the razor thin line between success and failure in the large and mega SC industry. OneWeb began launching its planned 650-unit large SC in 2019, however, in March 2020 the company filed for bankruptcy,[23] laying off the majority of employees but maintaining their satellite operations centre to manage the 68 units already in orbit. In its filing OneWeb cited the impact of the COVID-19 crisis, although industry insiders pointed to pressure from competing giants SpaceX and Amazon.[24] Despite this, the company filed with the FCC in May 2020 to expand their SC to a staggering

---

[21] Cal Poly CubeSat Laboratory, 'CubeSat Design Specification', 2020, https://static1.squarespace.com/static/5418c831e4b0fa4ecac1bacd/t/5f24997b6deea10cc52bb016/1596234122437/CDS+REV14+2020-07-31+DRAFT.pdf.

[22] Roberts, T., 'Space Launch to Low Earth Orbit: How Much Does It Cost?', 2020, https://aerospace.csis.org/data/space-launch-to-low-earth-orbit-how-much-does-it-cost/.

[23] Clark, S., 'OneWeb files for bankruptcy', *Spaceflight Now*, 2020, https://spaceflightnow.com/2020/03/30/oneweb-files-for-bankruptcy/.

[24] Alley, A., 'OneWeb goes bankrupt, satellite Internet firm blames Covid-19', *Data Center Dynamics,* 2020, www.datacenterdynamics.com/en/news/oneweb-goes-bankrupt-satellite-internet-firm-blames-covid-19/#:~:text=British%20satellite%20firm%20OneWeb%20has,the%20current%20Covid%2D19%20crisis.&text=In%20January%2C%20the%20firm%20was,deal%20but%20it%20fell%20through.

48.000 units, however this was revised in January 2021 to 6.372.[25] OneWeb eventually exited bankruptcy after being purchased by a joint U.K. Government and Bharti Global consortium and resumed launches in December 2020.[26]

The scale of the sum total of proposed large and mega SC is truly astounding. The successful launch of just three of the mega constellations (Starlink with 41.493 units, Guo Wang with 12.992 units and Project Kuiper with 3.236 units), accounting for over 80% of the planned total units, would dwarf the 2.787 total active satellites in orbit as of 31 August 2020 by a factor of 20.[27] Clearly, large and mega SC represent a step-change in human activity in space.

Given this, it is perhaps unsurprising to find that large and mega SC present major challenges to space regulation. In particular, the authors identify three major issues posed by large and mega SC: (1) radio frequency spectrum congestion, an issue that has previously attracted extensive study[28]; (2) interference with both ground-based and space-based astronomy, caused by LEO satellites reflecting light and broadcasting radio signals towards observation equipment; and (3) the production of debris as a result of both planned and unplanned events, for example during deployment, as a result of collisions and from end-of-life disposal procedures. This is not intended to be an exhaustive list of potential large and mega SC issues, but is instead intended to provide a variety of perspectives through which to examine regulatory approaches and loci[29] of regulation within the space sector.

Ensuring that each of these challenges are adequately and appropriately addressed is critical to ensuring the long-term sustainability and responsible use of space. As with any industry in which the rate of technological development, is outpacing the production of new regulation, the question that policymakers and stakeholders are beginning to ask is 'are current regulations adequate for the current context, or are they insufficient?' This question has been the subject of numerous reports and articles, with the consensus of academics, policymakers, state, and inter-state actors being that new regulation is needed to adequately regulate NewSpace.[30] However, this analysis has not been followed by widespread rigorous examination of which regulatory approaches are best suited for present and future space industry challenges, and which bodies and institutions are best-suited as loci within which to pursue this

---

[25] OneWeb, 'OneWeb Streamlines Constellation', 2021, www.oneweb.world/media-center/oneweb-streamlines-constellation.

[26] Foust, J., 'OneWeb emerges from Chapter 11 with new CEO', *SpaceNews*, 2020, https://spacenews.com/oneweb-emerges-from-chapter-11-with-new-ceo/.

[27] Union of Concerned Scientists, 'UCS Satellite Database', 2020, www.ucsusa.org/resources/satellite-database.

[28] See e.g., Tăiatu, C.M., 'The Future Impact of the ITU Regulatory Framework on Large Constellations of Satellites', in Froehlich, A. (ed.) '*Legal Aspects Around Satellite Constellations*', Springer, 2019.

[29] Within this article 'loci', plural of 'locus', is defined as locations or sites within which regulation can be pursued. Specifically, this refers to industry, national or international bodies capable of producing regulation.

[30] See e.g., O'Sullivan, S., 'NewSpace must be regulated', *SpaceNews*, 2019, https://spacenews.com/newspace-must-be-regulated/.

much-needed regulation. The authors aim to answer these questions, applying a conceptual framework of regulatory approaches to a general analysis of the various possible loci for regulation in order to identify the most appropriate such sites for addressing the challenges posed by large and mega SC.

In section two, the authors introduce their conceptual framework of regulatory approaches, discussing a variety of different factors which shape how a regulatory instrument functions. The choices of targeted versus generalised, rules-based versus principles-based, and hard law versus soft law, as well as combinations thereof, are explored in detail. This section is intended to illustrate the range of options available to policymakers when pursuing regulation, with the central argument that there is no one-size-fits-all approach, but that a bespoke solution with one or more tailored regulatory approaches can be highly effective.

Section three provides an overview of the various relevant actors and institutions active within this field that are potential loci within which to pursue regulation. The loci are split into three categories: industry level, in which self-regulation can be pursued by the private companies themselves; national level, in which state agencies or legislatures can produce nationally applicable regulation for any actor operating under its jurisdiction; and international level, in which coordination across and between states is possible. Loci within each level bring their own unique advantages and disadvantages, however, whilst each level has its own role to play within the broader regulatory ecosystem, the authors argue that loci within the international level are best suited for the pursuit of regulation of large and mega SC.

Finally, in section four the authors review each of the previously identified issues (congestion, interference, and debris), examining the specific nature of each problem, how this can inform the choice of appropriate regulatory approach, and which potential loci are best positioned to facilitate this regulation. The first of these, radio frequency spectrum congestion, has already begun to be addressed, with an initial solution having been implemented in 2019. Therefore, analysis of this issue is included primarily as an illustrative example to demonstrate how the approach-loci relationship the authors explore conceptually maps onto real-world issues.

2 Approaches to Regulation

The authors propose a conceptual framework of regulatory approaches rooted in three core regulatory design choices: (1) whether or not the regulatory instrument is targeted, exclusively applicable to large and mega SC, or generalised, applicable to a broader range of space activity; (2) whether it is principles-based, setting out principles of activity outcomes and leaving the means by which these goals are reached undefined, or rules-based, dictating the means but not necessarily the outcomes; and (3) whether it is hard law, binding and enforceable, or soft law, non-binding and non-enforceable. In proposing this conceptual framework, the authors build upon previous literature examining the character, functioning and relative suitability of principles-based vs. rules-based and hard law vs. soft law regulatory design.[31] Each of these choices warrants deeper discussion.

2.1 Targeted Versus Generalised

Among the first decisions necessary to make when pursuing regulation is its scope. This decision manifests as the choice of whether to adopt a targeted approach aimed at addressing the issue at hand as a large or mega SC specific concern, or a generalised approach aimed at introducing rules and principles for best practice for all space activity is more appropriate. Put simply, whether the problem should be treated as specific to large and mega SC or as a broader concern.

Although each of the issues under examination (congestion, interference, debris) are brought into sharp relief by large and mega SC, they are not necessarily all unique to this context. For example, production of space debris is a challenge common to any activity which involves launching objects into space. Granted, the scale of large and mega SC introduces new urgency to the problem, however, it remains an issue common all activities that involves launching objects into space. Conversely, interference with optical astronomy is an issue specific to large and mega SC.[32] It is the inherent and unique quality of large and mega SC, that they are comprised of 100s or 1.000s of satellites, that causes the issue, and therefore the issue cannot be separated from large and mega SC.

It follows logically that the regulation of large and mega SC specific issues is best achieved through a targeted approach, focusing explicitly only on this context, whilst generalizable issues would, on-balance, benefit from a generalised regulatory approach. Of course, the latter could still be regulated with a scope specifically limited to the context of large and mega SC; however, this is not necessarily desirable as in some cases the widely recognized need to regulate large and mega SC might afford policymakers a key opportunity to harness international consensus to address broader issues. For example, whilst the production of space debris could be tackled with a targeted large and mega SC specific approach, the issue may provide a valuable chance to produce generalised regulation on the production of space debris applicable to all space activities. In other words, if there is an opportunity to solve a broader problem, it is often advantageous to do so.

Of course, this is not always the case. Adopting a generalised regulatory approach inherently introduces more aspects to any given issue. By broadening the scope of the regulatory instrument's impact policymakers also invite a greater degree of

---

[31] See e.g. Abbott, K.W., Snidal, D., 'Hard and Soft Law in International Governance', *International Organization*, 2000, vol. 54(3), pp. 421–456; Schaffer, G., Pollack, M.A., 'Hard Versus Soft Law: Alternatives, Complements and Antagonists in International Governance', *Minnesota Law Review*, 2010, Vol. 94, pp. 706–99; Alexander, D., Jermakowicz, E., 'A true and fair view of the principles/rules debate', *ABACUS*, 2008, vol. 42(2), pp. 132–164.

[32] Technically it is possible for any space-based object to interfere with optical astronomy under specific conditions, however the scale of large and mega SC make them uniquely disruptive.

contestation over the proposed regulation itself. This is especially true where an issue touches on strategic concerns, as is the case with space debris and anti-satellite (ASAT) weapons systems. Additionally, generalised approaches require loci capable of introducing generalised regulation. Finding a body with a mandate to do so, especially within the context of space affairs, means looking primarily at international institutions such as United Nations (UN) bodies, in particular the UN Office for Outer Space Affairs (UNOOSA) and the Committee on the Peaceful Uses of Outer Space (COPUOS). These bodies, while ideal for examining broad issues, can be less capable of dynamic, responsive decision-making than more specialized organizations. As such, opting for a generalised regulatory approach where a targeted approach is possible can have the effect of slowing, or in extreme cases preventing, the successful production of any regulation whatsoever.

Nonetheless, it is the view of the authors that, where possible, a generalised approach to regulation should be prioritized over a targeted approach.

2.2 Rules-Based Versus Principles-Based

The second question facing policymakers is whether to pursue a rules-based or principles-based model of regulation. At its most basic, a rules-based approach dictates in detail a set of rules governing how actors may behave.[33] It seeks to achieve a given outcome by limiting the scope of actor behaviour, producing unambiguous applicable rules for actors to adhere to. Whilst rules-based regulation is both robust and easily measured for compliance, it is also rigid, leaving little room for innovation on the part of actors. Principles-based regulation inverts this approach, placing the emphasis on outcomes rather than process. Principles-based regulation prescribes principles for acceptable outcomes of actor behaviour but allows actors themselves to determine what measures and procedures are used to achieve that outcome.[34] This has the advantage of granting actors more leeway to innovate in their behaviours, whilst still regulating the actual impact of that behaviour.

Both rules-based and principles-based regulation have their relative advantages and disadvantages, and either may be better suited to a given issue. It is important to recognize that the space industry has a higher rate of change than many other sectors, with previous rules-based regulation being too rigid to allow regulatory frameworks to keep pace. Whilst it is not the case that a principles-based approach is always the more effective option, the relative flexibility of principles-based regulation offers a key advantage for a fast-growing industry that policymakers are keen not to stifle.[35]

---

[33] Conradie, L., 'Rules-based approach vs Principle-based approach to regulation in the Financial Industry', *Etude*, 2019, www.etude.co.za/article.php?article=32#:~:text=A%20rules%2Dbased% 20approach%20to,for%20each%20organisation%20to%20determine.

[34] Ibid.

[35] For a more in depth analysis of the relative advantages and disadvantages of rules- and principles-based regulation, see Decker, C., 'Goals-based and rules-based approaches to regulation', U.K.

Consider the issue of interference with astronomical observations. A rules-based approach may be to identify materials or coatings for satellites that reflect an acceptably low amount of light back to the Earth, and to dictate that all new LEO satellites must be made using these materials/coatings. This has the benefit of being clear, simple to apply and easy to measure compliance with. In contrast, a principles-based approach may instead focus exclusively on the outcome, setting out an upper limit of light reflected to the Earth, but leaving the means by which that is achieved up to manufacturers and operators to decide. This can be harder to apply, but allows actors to find potentially novel, innovative solutions perhaps unrelated to the materials used in manufacturing.

Additionally, whilst it is true that rules-based regulation is often more precise and more easily applied than principles-based frameworks, it is wrong to assume that its application is always uncontested. Rules-based international treaties are applied through a process of interpretation, especially in contexts where the law is not perfectly designed to match the situation it is being applied to.[36] The language of treaties is often vague and undefined, intentionally written so as to leave sufficient scope for actors to interpret and apply the law in different ways. A prime example of this is the *'Treaty on Principles Governing the Activities of States in the Exploration and Use of Outer Space, including the Moon and Other Celestial Bodies'*, commonly known as the *Outer Space Treaty* (OST), which prohibits the placement of any weapons of mass destruction in orbit, but leaves both 'weapons of mass destruction' and 'in orbit' undefined.[37]

Both rules-based and principles-based frameworks can be more or less appropriate than the alternative depending on the requirements of the situation. If regulating a highly specific technical challenge, or in cases where uniform behaviours are required to ensure a desired outcome, it may be more appropriate to set out clearly applicable rules for actors to follow. However, if the issue at hand is more general, or if a given outcome is required but not necessarily with uniform behaviour on the part of all actors, principles-based regulation may be more suitable.

2.3 Hard Law Versus Soft Law

Finally, policymakers must choose between either hard law or soft law, labels used to refer to the enforceability of the regulatory instrument.

Government, Department for Business, Energy & Industrial Strategy, 2018, www.econstor.eu/bitstream/10419/196215/1/2018-08-regulation-goals-rules-based-approaches.pdf.

[36] See Hurd, I., *'How to Do Things with International Law'*, Princeton University Press, 2017.

[37] United Nations, 'Treaty On Principles Governing The Activities Of States In The Exploration And Use Of Outer Space, Including The Moon And Other Celestial Bodies', 1967, A/RES/2222(XXI), www.unoosa.org/oosa/en/ourwork/spacelaw/treaties/outerspacetreaty.html.

Hard law, meaning binding legal regimes with robust enforcement mechanisms,[38] is often considered the 'gold standard' for international regulation. It imposes enforceable obligations and standards on actors subject to it, sanctioning actors who ignore or otherwise breach these obligations with punitive measures. Prime examples of hard law are international treaties, such as the aforementioned OST. Conversely, soft law, being non-binding and non-enforceable,[39] is often perceived as less robust, less able to compel actor behaviour, and therefore less valuable for regulation. However, whilst hard law certainly carries with it the key advantage of ensuring actor compliance via enforcement mechanisms, there are several reasons why pursuit of a new hard law framework is not always the most appropriate, realistic, or desirable approach.

Soft law is, broadly speaking, far easier to achieve than hard law.[40] Many actors, whether states concerned with the preservation of their sovereignty or companies seeking to maximize their profit margins, are reticent to subject themselves to binding measures which constrain their ability to act as they see fit. Negotiations of proposed binding international law are often highly contested and can involve lengthy discussions and undesirable compromises in order to secure an agreement. Within the context of large and mega SC, states with an active interest in space commercialization (e.g., the US) may fear losing a competitive edge in relation to other states and are primed to view binding regulation as stifling development and discouraging commercial actors from choosing to do business in that state.[41]

This is not to say that hard law is unachievable. The bedrock of the first generation of space regulation is formed by the five key space treaties: the previously mentioned OST (1967); *'Agreement on the Rescue of Astronauts, the Return of Astronauts and the Return of Objects Launched into Outer Space'* (1968); *'Convention on International Liability for Damage Caused by Space Objects'* (1972); *'Convention on Registration of Objects Launched into Outer Space'* (1976); and *'Agreement Governing the Activities of States on the Moon and Other Celestial Bodies'* (1984).[42] Space has always been a pioneering area of international cooperation; it is not entirely unreasonable to suggest that concerted international effort may yield new treaty law governing contemporary space activities. Nonetheless, the fact remains that producing new treaty law could be a drawn-out process.

Soft law, in comparison to hard law, refers in this context to any non-binding framework. Often, such frameworks will be focused on establishing principles of

---

[38] ECCHR, 'Hard Law/Soft Law', www.ecchr.eu/en/glossary/hard-law-soft-law/#:~:text=Hard% 20law%20refers%20generally%20to,legally%20enforced%20before%20a%20court.

[39] Ibid.

[40] Schaffer, G., Pollack, M.A., 'Hard Versus Soft Law: Alternatives, Complements and Antagonists in International Governance', *Minnesota Law Review*, 2010, Vol. 94, pp. 706–99.

[41] See e.g. Dr. Scott Pace (then Deputy Assistant to the President and Executive Secretary of the National Space Council) in discussion on this subject (Space Policy Pod, 'Episode 01: A Space Policy Discussion with Scott Pace', 2020, https://open.spotify.com/episode/2lljL5VVHGjMUBm fgUMmLB?si=Bk3oGTOiSAWFsW4CZ1KHCA).

[42] See UNOOSA, 'Space Law Treaties and Principles', www.unoosa.org/oosa/en/ourwork/spa celaw/treaties.html.

best practice which actors are encouraged to adopt themselves. As a result, soft law is often discussed exclusively in terms of principles-based regulation, with hard law conversely becoming synonymous with rules-based frameworks, however this is not necessarily accurate. Rules-based frameworks can be non-binding soft law, just as principles-based frameworks can be binding hard law. The difference between hard and soft law is that, whereas hard law compels actor behaviour through the threat of punitive measures, soft law only "expresses a preference and not an obligation that [an actor] should act, or should refrain from acting, in a specified manner."[43]

Nonetheless, it is true that soft law regulation is often principles-based. At the intersection of these two factors are norms. Norms are constraints, guidelines, or directives of actor behaviour as well as being products of that behaviour.[44] Norms are formed when actors, either unilaterally or multilaterally, act in a certain way, according to a certain principle, without any binding rules compelling them to do so. They can then be formalized through non-binding declarations and regulatory frameworks, although this is not strictly speaking necessary.[45] Within the context of regulatory design for space activities, norms can be established within soft law frameworks as principles of best practice, such as with the UNOOSA Space Debris Mitigation Guidelines.[46]

Norms, and by extension soft law as a regulatory approach, are often dismissed by realist scholars and policymakers as being unenforceable and therefore ineffective; however, this is an overly simplistic view. Although technically unable to compel actor compliance, norms and other soft law regulatory instruments can nonetheless have a tangible impact on actor behaviour. Beyond the positive incentive for actors to abide by soft law in order to be seen as legitimate,[47] soft law can also be formally adopted within binding national legislation. Doing so translates a non-binding framework into a binding one able to compel non-state actors operating within that nation's jurisdiction to comply. Whilst the Space Debris Mitigation Guidelines are non-binding soft law, they have been adopted either in their entirety or in part by a number of states.[48] Finally, if a soft law framework is sufficiently supported through actor behaviour and opinio juris, it can be considered customary international

---

[43] Gold, J., 'Interpretation: *The IMF and International Law'*, London, The Hague, Boston: Kluwer Law International, 1996: 301.

[44] Axelrod, R., 'An Evolutionary Approach to Norms', *The American Political Science Review*, 1986, vol. 80(4), pp. 1095–1111, www.jstor.org/stable/1960858?seq=1.

[45] See e.g. the French Ministry of Defence's publication on the application of international law to cyberspace (Roguski, P., 'France's Declaration on International Law in Cyberspace: The Law of Peacetime Cyber Operations, Part I', *OpinioJuris*, 2019, https://opiniojuris.org/2019/09/24/frances-declaration-on-international-law-in-cyberspace-the-law-of-peacetime-cyber-operations-part-i/.

[46] UNOOSA, '*Space Debris Mitigation Guidelines of the Committee on the Peaceful Uses of Outer Space*', 2010, www.unoosa.org/pdf/publications/st_space_49E.pdf.

[47] Foot, R., Walter, A., 'Global norms and major state behaviour: The cases of China and the United States', *European Journal of International Relations*, 2012, vol. 19(2), pp. 329–352, https://journals.sagepub.com/doi/abs/10.1177/1354066111425261.

[48] See e.g., UNOOSA, 'Compendium of space debris mitigation standards adopted by States and international organizations', www.unoosa.org/oosa/en/ourwork/topics/space-debris/compendium.html.

Table 1 Conceptual framework of regulatory approaches

	Rules-based	Principles-based
Hard law	Binding rules frameworks e.g., ITU Radio Regulations	Binding principles frameworks e.g., OST
Soft law	Non-binding rules frameworks e.g., ISO Standards	Non-binding principles frameworks e.g., UNOOSA Space Debris Mitigation Guidelines

law, universally legally binding upon all actors.[49] Soft law should not be dismissed out of hand.

Embedded in the choice of hard law or soft law is a trade-off between the ease of establishing the regulatory framework and how far that framework imposes enforced, binding obligations on actors. This trade-off will necessarily need to be evaluated within the context of the unique challenges posed by large and mega SC.

2.4 Conceptual Framework and Multiple Instruments Approach

Considering the factors discussed above, it is now possible to consider the variety of regulatory approaches available for policymakers addressing congestion, interference, and debris. Table 1 displays the various possible combinations of regulatory approach, with real-world examples of each included.

The question of targeted versus generalised regulatory solutions is notably absent from this framework, as the choice primarily affects the content of the regulation rather than its functional design. Any of the above combinations could be used to produce both targeted and generalised regulatory measures.

Despite the range of regulatory approaches available to policymakers, and the diversity of characteristics each offers, it is highly plausible that any single approach is simply insufficient to adequately regulate a given issue. This may be because an issue is broad, with too many distinct aspects to it to be solved through any one approach, or due to the disadvantages of that single approach being too great to accept without any additional measures being taken to mitigate them.

Thus, multiple simultaneous efforts with multiple differing approaches resulting in multiple complementary regulatory instruments may in some cases be the most effective strategy for regulators concerned with complex issues. In doing this, policymakers can tackle different aspects of a single broader issue without having to compromise on any of the advantages of different regulatory approaches. An example of this concept is that while a generalised principles-based soft law framework, the

---

[49] Kunz, J.L., 'The Nature of Customary International Law', *The American Journal International Law*, 1953, vol. 47(4), pp. 662–669, www.cambridge.org/core/journals/american-journal-of-intern ational-law/article/abs/nature-of-customary-international-law/330AC990F38C7CA7163596A9 D4E95358#.

Space Debris Mitigation Guidelines, has been an effective way of establishing principles of best practice, this could be supplemented or built upon by producing targeted rules-based hard law in the form of binding technical or data sharing requirements for automatic collision detection and avoidance systems. There is no one-size-fits-all approach, and no arbitrary need for only one solution.

3 Loci of Regulation

Appropriate choice of regulatory approach is the cornerstone of any effective regulation; however, it must be supported by a corresponding selection of locus within which to pursue the regulation. In the following section, the authors review potential loci of regulation within three distinct tiers: (1) industry level regulation, or loci for self-regulation by commercial actors (e.g., SpaceX); (2) national level regulation, both within national space agencies and at the executive level; and (3) international level regulation, consisting both of ad hoc multilateral initiatives and international bodies.

As with regulatory design, industry level, national level, and international level loci each offer their own advantages and disadvantages as sites for potential regulation, and within these imperfect groupings individual institutions and bodies can vary from one another significantly, resulting in a wide range of possible options for policymakers to select from.

To a certain degree, the choice of locus can be informed by the desired regulatory approach. Certain loci are inherently better suited to certain regulatory approaches, with technical competence, actors involved, decision-making level and scope of impact all contributing to this variance. Actors can therefore choose to pursue regulation in whichever locus is best suited to their approach.[50] Whilst the structure of this article implies that a choice of regulatory approach proceeds the choice of locus, this is not necessarily a unidirectional relationship. The choice of locus can often be restricted through other factors, for example, whether a given locus is available within a given time frame or whether all relevant parties are represented within it.

Just as there is no inherent reason that an issue should only have one regulatory solution, there is also no inherent reason that any single locus is the *only* appropriate site for regulating that issue. An effective regulatory strategy may involve multiple concurrent regulatory efforts occurring within multiple different loci.

---

[50] This is true with the exception of when the actor in question has no representation within a given institution, or where that institution is otherwise unable to facilitate this.

3.1 Industry and Self-regulation

Can the space industry regulate itself? When it comes to large and mega SC the industry is currently dominated by private companies motivated primarily by profit and growth. The drive to secure profit, and to not be left behind when others do, pushes commercial actors to maximize advantages and minimize risks to their business. This, especially where exacerbated by a lack of appropriate regulation, can quickly lead to commercial practices that are at best unsustainable and at worst predatory.

Yet, it is also true that industry self-regulation can, under certain circumstances, be an effective system. Within the context of the present challenges there is one international body particularly well-suited as a locus for industry self-regulation. The International Organization for Standardization (ISO) is an international standards-setting organization with a membership of 165 national standards organizations.[51] The ISO develops industry standards within any one of over 250 technical committees, covering everything from rivets to packaging dimensions, cosmetics to sustainable finance. Technical Committee 20, Aircraft and Space Vehicles, Subcommittee 13, Space data and information transfer systems (TC20/SC13 or SC13), and Subcommittee 14, Space Systems and Operations (TC20/SC14 or SC14), are responsible for space industry standards.[52]

SC13 and SC14 are comprised of representatives of the national standards agencies of twelve and 14 states respectively, with the former having the direct responsibility for 86 published standards with 18 currently in development, the latter 175 with 49 in development. ISO standards are non-binding and as such are soft law rather than hard law. However, adherence to ISO standards carries a great deal of legitimacy, and companies may request to become ISO certified in order to improve their perception, which includes adherence to ISO standards. ISO certification is a coveted status and one that is pursued by almost every major actor within the space industry, notably including SpaceX, and thus the ISO standards are in a certain sense 'self-enforcing'.

Thus, despite technically being soft law, in practicality they are closer to hard law in terms of actual compliance. As such, the ISO offers policymakers an effective locus within which to pursue soft law rules-based—ISO standards are a rules-based framework—regulation. ISO standards can be both targeted at specific types of spacecraft, or applicable to space activity more generally, making the ISO capable of both targeted and generalised regulation.

ISO standards are only developed upon request from industry actors or other stakeholders such as consumer groups, with companies communicating the need for an industry-wide standard to their respective national agency who may then

---

[51] ISO, 'About Us', www.iso.org/about-us.html.

[52] ISO, 'ISO/TC20/SC13 Space data and information transfer systems', www.iso.org/committee/46612.html; ISO, 'ISO/TC20/SC14 Space systems and operations', www.iso.org/committee/46614.html.

pursue the standard on their behalf within SC13 or SC14.[53] The actual development of the standard is done in conjunction with industry, consumer, governmental, NGO and academic input, with the bulk of experts working within the industry. As such, industry has a leading voice in both deciding which standards are pursued and developing the standard itself, giving industry a means through which to self-regulate.

On its own, the codification of technical standards is a powerful tool available to policymakers. Minimum standards for automatic collision detection and avoidance, data sharing, redundancies could play a valuable role in avoiding the production of space debris, and standards dictating the maximum reflectivity of satellite materials could aid in reducing interference with astronomical observation. Nonetheless, self-regulation is most effective when it plays a role within a broader regulatory ecosystem, rather than as the only regulatory system altogether. By itself, it is far less desirable.

The main issue with self-regulation in isolation is that it is demonstrably ineffective and inefficient. During September 2019, a potential collision between a European Space Agency (ESA) operated satellite and a SpaceX Starlink satellite prompted ESA to reach out to SpaceX to warn the company. SpaceX dismissed the original 1 in 50.000 risk as not requiring any action. When the risk of collision lowered a week later to just 1 in 1.000 (far closer than ESA's own 1 in 10.000 risk threshold), SpaceX did not respond to ESA's follow up communications, citing "a bug in our on-call paging system".[54] Perhaps the most widely cited and frankly definitive example of the failure of industry self-regulation is that only 30% of satellite operators are currently adhering to guidelines requiring satellites to be deorbited after 25 years.[55]

Self-regulation, while a potentially important piece of the evolving regulatory environment, cannot be the only route through which regulation is pursued. This notwithstanding, the ISO holds significant promise as a locus of regulation, not least for its ability to produce what are for all intents and purposes binding rules-based targeted and generalizable standards without having to engage in the often highly contested process of securing new binding international agreements.

3.2 National Bodies and Institutions

At first glance, national level regulation for an international issue appears to be self-evidently inadequate. Large and mega SC do not exist within the physical boundaries of any single state, nor is space owned or administered by any single state. Therefore, no single state can unilaterally introduce regulation to solve the issues raised by these constellations. Unlike industry self-regulation which can, through international

---

[53] See ISO, 'Developing Standards', www.iso.org/developing-standards.html.

[54] Minter, A., 'Viewpoint: It's Time to Regulate Outer Space', *Claims Journal*, 2019, www.claimsjournal.com/news/international/2019/09/11/292992.htm.

[55] Krag, H. reported in Foust, J., 'Starlink failures highlight space sustainability concerns', *SpaceNews*, 2019, https://spacenews.com/starlink-failures-highlight-space-sustainability-concerns/.

organizations like the ISO, incorporate all actors regardless of national affiliation, national level regulation is only applicable to actors operating within that state. Nonetheless, the companies that manufacture, launch, and operate large and mega SC are registered in states, and are subject to their parent states' laws. Therefore, it is worth examining the role that national loci can play in the pursuit of regulation.

As per the OST, states are responsible for the actions of non-state entities operating under their jurisdiction.[56] This means that, for example, the U.S. is internationally responsible for the actions of SpaceX, and therefore for the Starlink mega SC. Thus, the U.S. has an incentive to ensure that Starlink operates according to applicable international law. In effect, international law is therefore internalized and made national. This process of internalization is not unique to space activity. On a technical level, much of international hard law requires national approval before it applies to a given state; for example, international treaties require ratification in the U.S. Senate before they become binding.

This process may sound like a formality, and in many cases it is, however sufficiently motivated states can and sometimes do decline to take this final step. The U.S. is particularly notable, having neglected to ratify the 1989 *Convention on the Rights of the Child,* the 1997 *Ottowa Treaty* (landmine ban), and the 2002 *Optional Protocol to the Convention Against Torture,* among many more.[57] Given the importance of the U.S. within the present context (with many large and mega SC companies being US-based), this trend may indicate that seeking international regulation without U.S. national support may be ineffective, implying that actively pursuing regulation within U.S. national loci may be a valuable step towards securing international regulation.

The internalization of international regulation is not limited to hard law, it can offer a path towards internalization of non-binding principles and norms. The adoption of international soft law, for example the Inter-Agency Space Debris Coordination Committee (IADC) Space Debris Guidelines, directly within national legislation introduces a formal mechanism through which non-binding norms become binding law, transforming international soft law into national hard law. This process can also be indirect, such as national law requiring that actors abide by industry best practices being interpreted as compliance with ISO standards or the IADC guidelines. Thus, internalization, pursued within national loci, could be a particularly promising option within the present context, especially in cases where the major regulatory failing is not that no regulation exists, but that it is non-binding and not adequately complied with.

States may also unilaterally pursue regulation on their own initiative. Examples include Australia's Space Activities Act (1998), the Russian Federation's numerous Presidential Edicts and Decrees, the UK's Outer Space Act (1986), and the US's

---

[56] United Nations, 'Treaty On Principles Governing The Activities Of States In The Exploration And Use Of Outer Space, Including The Moon And Other Celestial Bodies', 1967, A/RES/2222(XXI), www.unoosa.org/oosa/en/ourwork/spacelaw/treaties/outerspacetreaty.html.

[57] See e.g. Human Rights Watch, 'United States Ratification of International Human Rights Treaties', 2009, www.hrw.org/news/2009/07/24/united-states-ratification-international-human-rights-treaties.

United States Code Title 51.[58] The U.S. has been particularly active throughout the Trump administration, with the revival of the National Space Council in 2017 catalysing the publication of four major Space Policy Directives (SPD) aimed at stimulating new development of U.S. national and commercial space activities.[59] In particular, SPD-2, *'Streamlining Regulations on Commercial Use of Space'*, directs national agencies to review and revise national regulations to regulate the needs of the contemporary space industry more appropriately.[60]

Insofar as national bodies for the development of space policy specifically, or agencies for the governance of space activities more generally, exist within any given state, they can be key loci for securing national regulation, whether through the internalization of international regulation or through the unilateral production of new policy. As specialized bodies aligned behind a common value-set and shared goal (set by the national executive), they can be highly responsive and quick to address new regulatory concerns. Additionally, while these loci often do not have a legislative mandate (e.g., NASA cannot write laws), their expertise is usually respected by legislative bodies, meaning that they have a key voice in the production of national level hard law.

Unfortunately, however, a common result of unilateral legal efforts is a wide variety of legislation that increasingly diverges from other states' laws—a fragmentation of space law into numerous different directions. Fragmentation introduces future difficulties for seeking international regulation, as regulatory alignment becomes more difficult to achieve. Furthermore, if any given state introduces new national legal frameworks to regulate space activities, these frameworks will only be applicable to that state. A state which unilaterally imposes additional constraints on its space industry risks becoming a less attractive prospect for commercial actors than states without those constraints, making it less competitive.

Therefore, finding internationally applicable, or universal standards of regulation is critical. By avoiding creating an uneven playing field for states seeking to attract and develop commercial space industry, attaining buy-in from the most active states becomes much easier. Additionally, this avoids the potential issue of 'flag of convenience' business practices, in which an actor registers their assets (SC) in a state with fewer constraints than others.

Although it is possible for national loci to play a key role in producing international outcomes, it is usually more effective to develop international standards in international loci. Of course, this is in some ways a false dichotomy, as national bodies and institutions are often the primary actors within international and multilateral organizations anyway. This is particularly evident in this context, as numerous national space agencies are members of, and act through, the IADC and other such

---

[58] UNOOSA, 'National Space Law Collection', www.unoosa.org/oosa/en/ourwork/spacelaw/nationalspacelaw/index.html.

[59] Space Foundation Editorial Team, 'Space Briefing Book: U.S. Space Laws, Policies and Regulations, U.S. Government', www.spacefoundation.org/space_brief/space-policy-directives/.

[60] White House, 'Space Policy Directive-2, Streamlining Regulations on Commercial Use of Space', 2018, www.whitehouse.gov/presidential-actions/space-policy-directive-2-streamlining-regulations-commercial-use-space/.

organizations. Ultimately, the primary issue with national level regulation is that management of space activity is an international concern, requiring high degrees of international coordination and regulatory alignment for effective regulation. It is not that there is no path from national loci to international regulation, but that there exist better suited loci within the international sphere which can produce international regulation directly.

3.3 International Bodies and Institutions

Opting to pursue regulation directly at the international level has a number of critical long-term advantages. As previously noted, it is logical to pursue international regulation for international issues. Doing so not only ensures that no actors are able to bypass the regulatory instrument, but also creates a level-playing field within the industry, alleviating the concerns of some states which may otherwise be unwilling to regulate their national space industries. Moreover, international regulation is, generally speaking, more robust than national level regulation. Whereas governments change, priorities and values shift, and states may amend or nullify existing legislation, this is far more difficult within an international context.

A natural starting point for international governance is the UN. UNOOSA, responsible for the organization and functioning of almost all UN space related initiatives, is the most obvious point of departure for any efforts seeking international regulation of space activities. UNOOSA oversees UN programs and activities relating to, amongst others, capacity building, long-term sustainability of space activities, global navigation satellite systems, space treaty implementation and space debris.[61] As such, an international overview of the challenges posed by large and mega SC is clearly within UNOOSA's mandate.

Among UNOOSA run bodies and initiatives are the Programme on Space Applications (PSA), the UN Platform for Space-based Information for Disaster Management and Emergency Response (UN-SPIDER), the International Committee on Global Navigation Satellite Systems (ICG), the inter-agency UN-Space, and the recently launched World Space Forum (WSF) event series.[62] However, UNOOSA's most notable function is to act as the Secretariat for COPUOS, the permanent structured Committee established to facilitate international cooperation in the development of outer space.[63] COPUOS has two subcommittees, the Scientific and Technical Subcommittee (STSC) which discusses questions related to the scientific and technical aspects of space activities (including sustainability of outer space activities) and the Legal Subcommittee (LSC) which discusses legal questions relating to the

---

[61] UNOOSA, 'About Us', www.unoosa.org/oosa/en/aboutus/index.html.

[62] UNOOSA, 'Our Work', www.unoosa.org/oosa/en/ourwork/index.html.

[63] UNOOSA, 'Committee on the Peaceful Uses of Outer Space', www.unoosa.org/oosa/en/ourwork/copuos/index.html.

exploration and use of outer space.[64] Both subcommittees meet each year prior to the annual COPUOS plenary session.

The STSC, LSC and COPUOS plenary are well-suited loci within which to seek international regulation. As what is effectively the highest-level international forum, COPUOS is the best suited body for facilitating high level discussions aimed at finding international consensus. Furthermore, due to the seniority of both the organization and the representation within it, COPUOS provides a key opportunity to achieve actual results. The regularity of these meetings is also a major benefit, offering frequent and recurring opportunities for discourse. Finally, COPUOS's mandate as the 'forum for the development of international space law' makes it the natural locus for any international regulation of space activities. Indeed, COPUOS was instrumental in the creation of the five key space treaties.[65]

This is not to say that all issues should be solely addressed within COPUOS. In many cases other, more specialized, loci exist that could be more appropriate for regulating a particular issue. Nonetheless, COPUOS is likely to play at least a supporting role in any notable international regulatory initiative, whether through the production of norms upon which specialized organizations can build hard law, or through facilitating discussion, research, and negotiation on contested issues. It is in some ways the 'default' locus, regardless of regulatory approach.

Remaining within the UN but moving beyond UNOOSA, ITU has a central role to play in space industry regulation. The ITU predates the UN itself, having been established in 1865 as the International Telegraph Union with a mandate to connect telegraph networks between states.[66] Since then, its mandate has broadened with the advent of new communications technologies, and now includes competency for assigning satellite orbits and radio frequency spectrum assignments. It is also a regulatory agency, maintaining the universally binding 'Radio Regulations' (RR), a hard law framework governing a variety of subjects, ranging from the bringing-into-use (BIU) procedures for new satellites and their spectrum assignments to technical requirements for ground stations and receivers.[67]

The RR framework is effectively a rules-based hard law framework, covering both targeted and generalised rules, meaning that the ITU is a potential locus for pursuing this regulatory approach. However, what makes the ITU a particularly promising locus for regulation is the in-built structured adaptation mechanism, the World Radiocommunication Conferences (WRC). The WRC bring together states, international organizations, and various non-state stakeholders such as commercial telecommunications companies every three or four years to review and revise the RR, ensuring that required changes are made to keep up with societal and technical

---

[64] Ibid.

[65] Ibid. It is important to note that while COPUOS would be the natural starting locus for actors seeking to produce new international hard law, such as a treaty on space debris, this would ultimately need to be proposed, debated, and voted on in the form of a resolution at the annual UN General Assembly (UNGA).

[66] ITU, 'About International Telecommunication Union (ITU)', available www.itu.int/en/about/Pages/default.aspx.

[67] ITU, 'Radio Regulation', www.itu.int/pub/R-REG-RR.

developments.[68] This means that, through the WRC, the ITU can introduce new universal rules-based hard law on a regular basis.

The ITU has begun to address the unique challenges posed by large and mega SC. At WRC-19 the RR were revised to create new BIU regulations for large and mega SC, with a milestone-based approach, in which 10% of an SC's satellites must be launched within two years of the initial seven-year registration period finishing, updating an outdated framework that would have allowed companies to register an entire 1.000+ mega SC as having been deployed with the launch of a single satellite.[69] The ITU and the WRC-19 will be discussed in more depth in section 4.1; however, as an adaptable, permanent, structured forum for the regulation of telecommunications it clearly has the mandate, technical competency, experience and means to play a major role in future regulation.

Beyond the UN there are numerous other international bodies relevant to the specific concerns raised by large and mega SC. In particular, the IADC, ISO, and the Consultative Committee for Space Data Systems (CCSDS) merit further discussion.

The IADC is an international forum established for the coordination of activities related to issues of man-made and natural debris in space. Its constituent members are the national space agencies of 13 states, including the US's NASA, the China National Space Administration (CNSA) and the Russian Federal Space Agency (ROSCOSMOS), as well as ESA.[70] The primary purpose of the IADC is to provide its various members a forum through which to coordinate their activities on space debris mitigation. Although it has no formal legislative mandate, the IADC is a proven locus for the pursuit of soft law, particularly norms and other principles-based frameworks.

The IADC is best known for having published the first Space Debris Mitigation Guidelines in 2001, with subsequent revisions in 2007 and 2019. As previously noted, the guidelines are non-binding, however, they have been internalized within the national legislation of several states. More recently the IADC has begun working on the challenges posed by large and mega SC, having published an initial 'Statement on Large Constellations of Satellites in LEO' in 2019, and 'Potential Additional Mitigation Measures to Address the Proliferation of Small Satellites and Large Constellations' (circulating internally during 2020).[71] Thus, not only is the IADC a perfectly suited locus for soft law regulation of space debris, having been designed for precisely that purpose, it has already begun to act within the context of large and mega SC.

Having previously discussed the ISO at some length in the context of industry self-regulation, it is necessary only to reiterate that it is an international organization

---

[68] ITU, 'World Radiocommunication Conferences (WRC)', www.itu.int/en/ITU-R/conferences/wrc/Pages/default.aspx.

[69] Henry, C., 'ITU sets milestones for megaconstellations, *SpaceNews*, 2019, https://spacenews.com/itu-sets-milestones-for-megaconstellations/.

[70] IADC, 'Member Agencies List', www.iadc-home.org/member_agencies_list.

[71] Kim, H.D., 'IADC Statement on Large Constellations of Satellites in Low Earth Orbit', 2019, www.iadc-home.org/documents_public/view/page/2/id/111#u.

capable of producing industry wide rules-based regulation. SC13 and SC14 are effective loci for pursuing this regulation within the space sector, with the awareness that although ISO standards are non-binding, the widespread desire for ISO certification within the space industry does mean they are effectively self-enforced.

Similar in its constituent members to the IADC and in its standards setting function to the ISO, CCSDS is a multinational forum made up of eleven national space agencies which produces standards for space communications and data systems.[72] The organization produces standards across six technical areas, including mission operations, onboard interface systems, and space link services. As with the ISO, these standards are non-binding, though they are widely adopted. As such, the CCSDS is suited for the pursuit of targeted rules-based regulation and could perhaps provide a valuable contribution if pursued to supplement other efforts.

This is by no means an exhaustive list of all the international bodies and institutions engaged in space activity regulation. Many regional and issue-specific bodies such as the Arab Satellite Communications Organization (ARABSAT), the Regional African Satellite Communications Organization (RASCOM), the European Organization for the Exploitation of Meteorological Satellites (EUMETSAT), the European Telecommunications Satellite Organization (EUTELSAT-IGO), the International Mobile Satellite Organization (IMSO), and the International Space Exploration Coordination Group (ISECG) have not been discussed. Each of these may have some potential as loci for regulation, however, within the present context they are less suited than the previously discussed institutions and bodies, and a full analysis of each is beyond the scope of this article.

4 The Issues, Proposed Solutions and Appropriate Loci

Turning now to the three issues—congestion, interference, and debris—in the following section the authors review the nature of these challenges, suggest proposed regulatory approaches and identify appropriate loci in which to pursue that regulation. Although in this section the discussion of each issue follows a structure of issue-approach-locus, this should not be taken as a blueprint for how regulatory decision making is undertaken. Such a linear decision-making process is not necessarily how regulation is pursued, should be pursued, or at times can be pursued.

4.1 *Radio Frequency Spectrum Congestion*

Radio frequency spectrum congestion is an issue that has, to a certain extent, already been appropriately addressed. As such, it will be included here only in brief as a case through which to examine the issue-approach-locus relationship.

---

[72]CCSDS, 'Member Agencies', https://public.ccsds.org/participation/member_agencies.aspx.

Among the most obvious of issues presented by large and mega SC is the sheer number of satellites being launched. The congestion of space, both physical congestion of LEO orbits and radio congestion of the frequency spectrum satellites use to communicate with ground stations on the Earth surface, is a key concern.[73] Due to satellites being limited to the microwave band for Earth-satellite communications, the frequency spectrum is perceived as a scarce or even finite resource. As more satellites are launched, the demand for specific frequencies rises, and as demand rises, the scarcity of frequencies increases in tandem.[74] Scarcity drives competition over frequency assignments, leading to inefficient and unsustainable practices designed to maximize each actor's resources. This may result in a phenomenon known as spectrum warehousing.

Spectrum warehousing is the practice of filing for a significant amount of frequencies to support a theoretical large or mega SC, which are then reserved within the ITU Master International Frequency Register (the MIFR), blocking other actors from requesting them. Prior to the ITU WRC-19, the RR allowed actors to then deploy a single satellite out of a proposed large or mega SC and register the entire frequency allocation as having been brought into use. In doing so, the actor could effectively reserve options on vast amounts of frequency, both allowing the actor to guarantee access to its allocated frequencies long before it was prepared to use them and blocking other actors from requesting those frequencies for themselves.

However, to prevent spectrum warehousing, the ITU RR were revised during WRC-19, establishing new BIU protocols for large and mega SC. The new RR adopt a milestone-based approach, ensuring that allocated frequencies continue to be reserved only if the operator can meet certain deployment thresholds on a specified timeline, for example 10% of an SC's satellites must be launched within two years of the initial seven-year registration period finishing.[75]

Reviewing this case through the issue-approach-locus lens reveals the extent to which the three aspects are related. The issue, inherent to large and mega SC (targeted approach needed), is that absent enforced (hard law approach needed), appropriate instructions (rules-based approach needed), actors may engage in harmful behaviour—spectrum warehousing. Adopting a targeted, rules-based hard law approach brings us to either the ITU or COPUOS. From there consideration of any other factors, in this case the fact that the issue relates specifically to the ITU RR BIU regulations and should therefore be logically solved by adjusting those same regulations, aids in identifying the most appropriate locus.

Whereas the case of radio frequency spectrum congestion is relatively straightforward, this is not the case for either interference or debris. As such, both benefit to

---

[73] Note that, whilst physical orbit congestion is a key concern, it is impossible to separate from the issue of debris, and so is covered within that section.

[74] See e.g., Tăiatu, C.M., 'The Future Impact of the ITU Regulatory Framework on Large Constellations of Satellites', in Froehlich, A. (ed.) *'Legal Aspects Around Satellite Constellations'*, Springer, 2019.

[75] Henry, C., 'ITU sets milestones for megaconstellations, *SpaceNews*, 2019, https://spacenews.com/itu-sets-milestones-for-megaconstellations/.

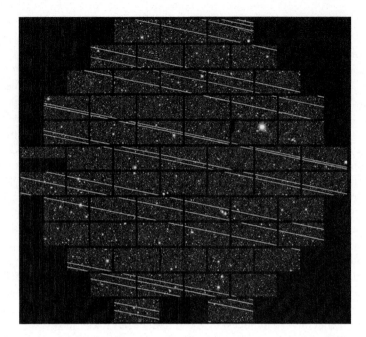

Fig. 1 Wide-field image from the dark energy camera on the Victor M. Blanco 4-m telescope at the Cerro Tololo InterAmerican Observatory, 18 November 2019. Several Starlink satellites can be seen in the image

a greater extent than congestion from a deeper consideration of regulatory approach and potential loci.

4.2 Interference with Astronomical Observation

An often-overlooked problem caused by the deployment of satellite constellations is interference with both ground-based and space-based astronomical observation. The light reflectivity of satellites and their solar arrays can cause light artefacts and errors on telescope sensors, or light trails for long-exposure captures, as seen in Fig. 1. For ultra-wide imaging exposures this is particularly concerning, as the saturation of the light trails left by bright satellites can have ruinous effects on the detectors and therefore compromise any observations made.[76] If the number of satellites in LEO continues to grow without proper mitigation errors, their adverse effects on scientific observation will severely hinder or even halt astronomical discoveries.

---

[76] Hainaut, O.R., Williams, A.P., 'Impact of satellite constellations on astronomical observations with ESO telescopes in the visible and infrared domains', Astronomy & Astrophysics, 2020, vol. 636(April 2020), www.aanda.org/articles/aa/full_html/2020/04/aa37501-20/aa37501-20.html.

On 25 August 2020, The SATCON1 Workshop published a report detailing the impact of SCs on optical astronomy, stating that existing and planned SCs in LEO will "fundamentally change astronomical observing at optical and near-infrared wavelengths".[77] Given that the projected surface density of satellites in LEO will be greatest near the horizon and during twilight, the report states that this will disproportionately affect scientific programs that require twilight observations, such as searches for Near-Earth Objects (NEO), distant Solar System objects, and optical counterparts to gravitational wave sources. The effect on other programs will range from negligible to significant and may still require novel software and hardware efforts to remove satellites and their trails from astronomical images.

The report goes on to state that satellites with an orbit below 600 km have the advantage of being hidden for several hours around local solar midnight due to entering Earth's shadow. However, their relative nearness to Earth means that the satellites are much brighter than if at higher orbital altitudes, ruining twilight captures for observatories at middle altitudes. The impact of satellites with orbital altitudes higher than 600 km are of even greater concern as they can be illuminated all night long. To put matters into perspective, recall that in May 2020 OneWeb sent a request to the FCC to augment the total number of satellites in their constellation to 6.372 at an orbital altitude of 1.200 km[78]; having such a large number of satellites illuminated throughout the night would render a large proportion of astronomical observation unusable.

The SATCON1 report provides several recommendations to both observatories and satellite operators to mitigate astronomical interference, including reducing the albedo of satellite surfaces, controlling satellite attitude to reflect less of the Sun, or—rather optimistically—launching fewer or no LEO satellites. The industry has seen efforts for self-regulation: in March 2020, a 60-satellite Starlink launch included a DarkSat prototype whose surface had been altered to reduce its albedo, with it reflecting 55% less light when in orbit compared to a normal Starlink satellite.[79] Although promising, the question remains about the long-term reliability of 'darkened' satellites, as the constant absorption of more heat than a reflective satellite is likely to incur a greater challenge in thermal management as the structure, avionics, electronics, fuel, and radiators of the spacecraft are placed under considerably more thermal stress.[80] Other solutions include sun visors that can cover highly reflective parts of a satellite and therefore reduce its reflectivity. In June 2020, SpaceX launched another prototype Starlink, dubbed 'VisorSat', that included such a visor; however,

---

[77] Walker, C., Hall, J. (eds.) *'Impact of Satellite Constellations on Optical Astronomy and Recommendations Toward Mitigations'*, American Astronomical Society, 2020, https://aas.org/sites/default/files/2020-08/SATCON1-Report.pdf.

[78] OneWeb, 'OneWeb Streamlines Constellation', 2021, www.oneweb.world/media-center/oneweb-streamlines-constellation.

[79] Foust, J., 'SpaceX claims some success in darkening Starlink satellites', *SpaceNews*, 2020, https://spacenews.com/spacex-claims-some-success-in-darkening-starlink-satellites/.

[80] This is an important aspect of thermal control within satellites and significantly narrows down the choice of paint for satellite exteriors.

astronomers are still awaiting further confirmation that the modifications applied by SpaceX will have a significant effect on a satellite's albedo.[81]

It is not clear if these steps taken by SpaceX are enough to mitigate optical astronomical interference—and even if they are, there currently exists no regulatory instrument other than the desperate plea of astronomers, professional and amateur alike, to avoid the damaging of astronomical observation. For now, companies must collaborate with various astronomical associations and observatories in order to discuss mitigation efforts. Leading these efforts is, once again, SpaceX. Shortly after the Astro2020 Decadal Survey Committee held an online conference in April 2020 on the topic of 'Astronomical Interference from Satellite Constellations', SpaceX posted an update to their website detailing previously undisclosed information about Starlink launches, orbit raises and parking orbits to elucidate plans for reduction of satellite reflectivity.[82] This includes rolling the satellite knife-edge to the Sun to minimize the reflection of light onto Earth during the orbit raise phase. During the satellite's on-station phase, the solar array is adjusted to be hidden behind the chassis of the satellite and the sun visors cover the highly reflective antennas of the satellite. SpaceX specifically states as a goal to have Starlink satellites in orbit with a magnitude of 7 (compared to the current 5.5); it is worth noting that a satellite of magnitude 7 brightness is still visible to the naked eye.[83] Even so, the fact remains that there is no obvious solution readily available.

This issue is currently specific to large and mega SC, though only due to the numbers being deployed. It is possible that in the future vast numbers of orbiting objects may be deployed for unrelated purposes, however, for the foreseeable future this issue exists as a direct result of an inherent quality of large and mega SC. As such, a targeted approach is sufficient. Whereas rules-based regulation would be an effective way of ensuring widespread adoption of a technical solution such as VisorSat's visor, it is unclear if this or any other presently identified technical solution is effective. Additionally, in prescribing a set of acceptable technical standards, either through the ISO or any other standards setting organization, regulation could prevent actors from seeking innovative solutions to the issue. Thus, until such a time as a working technical solution is identified, a principles-based approach is more appropriate. The exact content of such a regulatory framework would need to be debated at length by relevant stakeholders, however an option would be to establish a principle of minimal interference—a directive to all actors to take all possible steps to avoid unnecessary interference with optical astronomy through both design and operations solutions. Such a framework would preferably be hard law, binding upon all actors, however an initial soft law agreement could be a prerequisite stepping stone towards this goal.

---

[81] Witze, A., 'How satellite 'megaconstellations' will photobomb astronomy images', *Nature,* 2020, www.nature.com/articles/d41586-020-02480-5.

[82] SpaceX, 'Astronomy discussion with national academy of sciences', 2020, www.spacex.com/updates/starlink-update-04-28-2020/.

[83] See e.g., www.globeatnight.org/magnitudes.php; It is worth mentioning that the apparent magnitude scale is reverse logarithmic rather than linear; the larger the number, the dimmer the object. A difference of 1.0 in magnitude corresponds to a brightness ratio of around 2.5.

Within the international sphere, there are two loci within which policymakers might pursue regulatory solutions to this issue. Should an appropriate technical solution be identified, the ISO standards offer an ideal mechanism through which this solution could implemented industry wide. As regards a principles-based approach, COPUOS is the obvious choice. Capable of developing both hard and soft law, COPUOS has both the mandate and capability to facilitate the pursuit of regulation on this issue.

Astronomical interference is not limited to optical astronomy, also affecting radio observations. SC radio transmissions interfere with highly sensitive captures from arrays that normally operate in terrestrial radio-silent zones.[84] Starlink satellites send radio signals to ground stations on a frequency band running from 10.7 to 12.7 ghz, a range known as 5b, one of several such bands used by the Square Kilometre Array (SKA)'s South African site to search for simple proteins such as glycine.[85] An SKA internal analysis found that with only 6.400 units orbiting in LEO, the resultant radio transmissions would reduce the sensitivity of the 5b band by 70%. If the number of satellites increases to 100.000, the entire 5b band would be unusable.[86]

One potential solution is to build upon the existing concept of radio silent zones. This concept sees radio transmitters banned from certain areas within which their radio signals could interfere with observation equipment, for example the National Radio Quiet Zone (NRQZ) in the US. A hard law regulation compelling large and mega SC operators to coordinate with ground-based observatories to turn off transmissions from any unit orbiting above an astronomical observation site may provide a long-term solution. This concept is currently being explored by the U.S. National Radio Astronomy Observatory and SpaceX,[87] as well as within COPUOS run workshops examining both radio and optical interference more broadly.

The ITU is the clear choice of locus for either solution, able to build radio quiet zones into the hard law RR, ensuring actor compliance. However, the ITU is not scheduled to hold another WRC until 2023, which, at the time of writing, is over two years away. As an interim measure, it may be appropriate for the astronomy community to push for soft law guidelines on how to implement the radio quiet zones approach simultaneously pursuing either ISO standards or the aforementioned ITU RR changes. Given that radio observation occurs from sites located within the territory of a given state, this is a rare example of where national level regulation could be a viable, appropriate option. If each state implements its own radio quiet zone regulations to protect its radio observation sites then there may not be a need for international regulation of this issue at all.

---

[84] Clery, D., 'Starlink already threatens optical astronomy. Now, radio astronomers are worried', *American Association for the Advancement of Science,* 2020, www.sciencemag.org/news/2020/10/starlink-already-threatens-optical-astronomy-now-radio-astronomers-are-worried.

[85] Amos, J., 'Square Kilometre Array project frets about satellite interference', *BBC*, 2020, www.bbc.com/news/science-environment-54457344.

[86] SKA, 'SKAO Needs Corrective Measures From Satellite 'Mega-Constellation' Operators To Minimise Impact On Its Telescopes', www.skatelescope.org/news/skao-satellite-impact-analysis/.

[87] Green Bank Observatory, 'Joint NRAO and GBO Statement', 2019, https://greenbankobservatory.org/joint-nrao-and-gbo-statement/.

4.3 Space Debris

The production of space debris is likely the most recognized and pressing challenge facing the space industry. The UNOOSA Space Debris Mitigation Guidelines define space debris as "all man-made objects including fragments and elements thereof, in Earth orbit or re-entering the atmosphere, that are non-functional".[88] Thus, despite their impressive numbers, defunct satellites only make up a small portion of debris; rocket stages, launch hardware, solid propellant slag, deterioration fragments (such as peeling paint), fragments from exploding batteries or fuel tanks or from accidental or deliberate collisions are also encompassed within the definition.

Debris is a huge and growing problem. The number of debris objects regularly tracked by Space Surveillance Networks and maintained in their catalogue is 28.210; in comparison, it is estimated that 34.000 debris objects greater than 10cm are in orbit; 900.000 objects between 1cm and 10cm; and 128 million between 1 mm and 1 cm.[89]

To say that these uncontrollable objects are highly dangerous is an understatement. At the speeds reached in orbit, a satellite of CubeSat dimensions ($10 \times 10 \times 10$ cm, 1 kg) travelling at 7 km/s would have the kinetic energy equivalent of 6 kg of TNT.[90] Any collision between a satellite and any such debris could result in the loss of mission due to satellite hardware impact and, more importantly, the loss of the satellite and the creation of more space debris in orbit.

To date, there have only been two major debris-producing break up events in outer space, a 2007 Chinese anti-satellite (ASAT) weapons test which produced 3.037 new trackable pieces of debris,[91] and the 2009 collision between inactive Russian communications satellite Cosmos 2.251 and an active commercial communications satellite operated by US-based Iridium Satellite LLC, which produced an estimated 200.000 new pieces of debris.[92] The Cosmos-Iridium collision is a well-known case study within space policy, not least for the complex and unresolved liability issues raised by the event. The collision was not anticipated, with no central system in place to track objects and identify possible events.

Avoiding debris-producing collisions is a core concern of all space actors, however, the production of debris during launching, normal operations, by anticipated deterioration, and through insufficient end-of-life practices are also key causes that will become ever more important given the scale of the present generation planned large and mega SC.

---

[88] UNOOSA, 'Compendium of space debris mitigation standards adopted by States and international organizations', www.unoosa.org/oosa/en/ourwork/topics/space-debris/compendium.html.

[89] ESA, 'Space debris by the numbers', 2021, www.esa.int/Safety_Security/Space_Debris/Space_debris_by_the_numbers.

[90] Wittig, M., 'Space Debris and De-Orbiting', www.itu.int/en/ITU-R/space/workshops/2015-prague-small-sat/Presentations/MEW-Prague.pdf.

[91] Weeden, B., '2007 Chinese Anti-Satellite Test Fact Sheet', *Secure World Foundation*, 2010, https://swfound.org/media/9550/chinese_asat_fact_sheet_updated_2012.pdf.

[92] Weeden, B., '2009 Iridium-Cosmos Collision Fact Sheet', *Secure World Foundation*, 2010, https://swfound.org/media/6575/swf_iridium_cosmos_collision_fact_sheet_updated_2012.pdf.

Currently, there are no binding regulations governing the production of space debris. There are, however, several soft law frameworks addressing the issue. In particular, the IADC Space Debris Mitigation Guidelines, the subsequent UNOOSA Space Debris Guidelines, and the Space Debris Mitigation ISO Standards merit discussion.

The IADC Space Debris Mitigation Guidelines and the UNOOSA Space Debris Mitigation Guidelines (developed from the IADC framework), currently provide an international framework encouraging actors to avoid the production of debris. The IADC guidelines cover three main aspects: (1) limitation of debris released during normal operations, (2) minimization of the potential for on-orbit break-ups and collisions, and (3) removal of non-operational objects from populated regions.[93] Following the publication of the IADC guidelines, UNOOSA produced a separate framework, developed from the IADC regulations. This framework contains seven guidelines for actors to follow setting out principles of best practice regarding design, launch, operations, and end-of-life.[94] As with the IADC framework, these guidelines are non-binding, however they have been used as the basis upon which many national agencies have produced their own frameworks, such as ESA's 2008 Requirements on Space Debris Mitigation for Agency Projects.[95] The IADC and UNOOSA guidelines effectively provide a generalised soft-law principles-based framework of 'what' needs to be achieved in relation to space debris mitigation, leaving the 'how' these guidelines are implemented to national level hard law frameworks, international standards or actors' own discretion.

Foremost amongst international standards, the ISO Space Debris Mitigation standards offer high-level requirements (through ISO 24113) dealing with space debris mitigation requirements.[96] The standards were recently updated to provide stricter requirements for LEO orbits; the third edition of the documents,[97] released in 2019, included the following additional important changes worth discussing:

- Avoiding the intentional release of space debris into Earth orbit during normal operations by imposing a limit to the number of Launch Vehicle Orbital Stages (LVOS) to one for single-satellite launches, and two for multi-satellite launches; and ensuring that Solid Rocket Motors (SRMs) and pyrotechnics do not produce debris larger than 1mm into LEO or GEO.

---

[93] The IADC guidelines and subsequent revisions can be accessed on the IADC website. (IADC, '4 Main IADC Products', www.iadc-home.org/documents_public/view/id/82#u.

[94] UNOOSA, 'Compendium of space debris mitigation standards adopted by States and international organizations', www.unoosa.org/oosa/en/ourwork/topics/space-debris/compendium.html.

[95] UNOOSA, 'ESA Space Debris Mitigation for Agency Projects', www.unoosa.org/documents/pdf/spacelaw/sd/ESA_space_debris_mitigation_for_agency_projects.pdf.

[96] ISO 24113 is the high-level standard; lower-level standards include ISO 16127, 26872, 16164, 20893, 16699, 27875, 23339, 27852, 16126, 11227, 14200, as well as Technical Reports 16158, 18146, and 20590.

[97] ISO, 'ISO 24113:2019 Space systems—Space debris mitigation requirements', www.iso.org/standard/72383.html.

- Removing spacecraft and LVOS from protected LEO region after end of mission within 25 years from orbit insertion for spacecraft/LVOS without collision avoidance manoeuvres (CAMs), 25 years after end-of-life for entities with CAMs, or from the epoch of first intersection of an LEO region within 100 years after end of life if the spacecraft/LVOS operates outside the LEO region. The probability of disposal must be above 90%.
- The disposal option of manoeuvring to a higher orbit than the LEO region was removed, imposing a requirement for disposal accomplished by retrieval, controlled re-entry, natural orbit decay, or shortening orbit lifetime through manoeuvring or drag augmentation.

Although the ISO standards are technically non-binding soft law, they are widely adopted as formal requirements by most actors in the space industry. A notable example is the European Cooperation for Space Standardization (ECSS) via Adoption Notice of ISO 24113 in their Space Sustainability standard.[98] However, it was noted during the First International Orbital Debris Conference in 2019 that ISO 24113 and subsequent standards do not apply strict enough restrictions for the types of SCs that are currently being launched and proposed, suggesting that the standards will have to be updated in the near future to account for large and mega SCs.[99] This criticism is also true for both the IADC and UNOOSA guidelines.

In the case of the IADC, the organization itself began examining the impacts of large and mega SC in 2015, producing the 'IADC Statement on Large Constellations of Satellites in Low Earth Orbit' in 2017, an initial statement announcing that they themselves were aware of the inadequacy of the existing guidelines.[100] In particular, the IADC points to the end-of-life guidelines as a key area of concern, noting that the 25-year post mission disposal window, with a 10% accepted failure rate in disposal mechanisms, is insufficient within the context of large and mega SC.

This cause of this inadequacy relates to the problem of orbit congestion, which increases the risk of collisions as a result of the saturation of LEO. Critical object concentrations, at which there is an unacceptably high risk of collisions, are caused in part by satellite failure rates and the lack of end-of-life disposal of orbiting objects

The failure rates of current generation large and mega SC units vary from company to company. Starlink units have at different times had reported failure rates ranging from 5 to 0.2%.[101] At 40.000 units, this would mean somewhere between 80 and

---

[98] ECSS, 'ECSS-U-AS-10C Rev.1—Adoption Notice of ISO 24113: Space systems—Space debris mitigation requirements', 2019, https://ecss.nl/standard/ecss-u-as-10c-adoption-notice-of-iso-24113-space-systems-space-debris-mitigation-requirements-2/.

[99] Stokes, H., Akahoshi, Y., Bonnal, C., Destefanis, R., Gu, Y., Kato, A., Kutomanov, A., LaCroix, A., Lemmens, S., Lohvynenko, A., Oltrogge, D., Omaly, P., Opiela, J., Quan, H., Sato, K., Sorge, M., Tang, M., 'Evolution of ISO's Space Debris Mitigation Standards, *Journal of Space Safety Engineering*, 2020, vol. 7(3), pp. 325–331, www.sciencedirect.com/science/article/abs/pii/S2468896720300689.

[100] Kim, H.D., 'IADC Statement on Large Constellations of Satellites in Low Earth Orbit', 2019, www.iadc-home.org/documents_public/view/page/2/id/111#u.

[101] See e.g. Zafar, R., 'SpaceX Has Reduced Starlink Failure Rate to 0.2% Reveals Early Data', *WCCFTECH*, 2020, https://wccftech.com/spacex-starlink-failure-rate-early-data/; O'Callaghan, J.,

2.000 failed satellites in orbit. Not all of these will be propulsion failures, meaning that of these numbers a significant portion can be actively deorbited, however, at least some of these will not be actively manoeuvrable. SpaceX has planned for this, with defective units passively deorbiting and burning up in the atmosphere within a year without actively maintaining their orbit via propulsion.[102] However, not all large and mega SC are planned to be deployed to sufficiently low orbits for this deorbiting to occur naturally, or at a quick enough pace to keep up with launch rates.

With the prospect of a possible increase of mass in LEO of up to 90%, IADC member studies have found that a major criterion for maintaining the long-term sustainability of LEO is therefore the reliability of post-mission disposal of decommissioned satellites in accordance with existing debris mitigation guidelines.[103] The current guidelines (now also adopted within the ISO standards) dictate that end-of-life disposal rates must meet or exceed 90%, however the IADC noted that current compliance levels for LEO satellites that require an induced de-orbit have not reached 20% in any of the past 25 years.[104] Clearly, more needs to be done to ensure that actors cannot simply abandon their decommissioned or defunct satellites in space.

Evaluating this issue through the issue-approach-loci lens reveals, in terms of the conceptual framework of regulatory design, the inadequacies with current regulatory instruments. There already exist soft law principles-based frameworks (IADC and UNOOSA), supported by a rules-based framework which, while not inherently binding, is to a degree self-enforcing (ISO). However, both these rules and the proportion of actors adhering to them are insufficient. Therefore, two approaches are needed. First, the rules-based framework must be revised to ensure its adequacy in the context of large and mega SC. This can be achieved through further revisions to the ISO standards. Second, a universal binding requirement for compliance with both the rules-based and principles-based frameworks is necessary to ensure that all actors operate according to these regulations. Given the importance, scope, urgency, and severity of potential consequences, the authors argue that a generalised rules-based hard law framework, a 'space debris treaty', is the appropriate instrument through which to achieve this.

The proposed treaty would need to address the issue of space debris within three categories: (1) debris resulting from planned normal operations; (2) debris resulting from unplanned events; and (3) end-of-life object disposal.

Beginning with the first section, debris may be created during nominal operations of a launch vehicle and satellite operations. The most notable types of debris produced during the Launch and Early Orbit Phase (LEOP) come from pyrotechnics, launch

---

'Not Good Enough—SpaceX Reveals That 5% Of Its Starlink Satellites Have Failed In Orbit So Far', *Forbes,* 2019, www.forbes.com/sites/jonathanocallaghan/2019/06/30/not-good-enough-spacex-reveals-that-5-of-its-starlink-satellites-have-failed-in-orbit/?sh=689a1f2d7e6b.

[102] Henry, C., 'Contact lost with three Starlink satellites, other 57 healthy', *SpaceNews,* 2019, https://spacenews.com/contact-lost-with-three-starlink-satellites-other-57-healthy/.

[103] See e.g., Aloia, V., 'The Sustainability of Large Satellite Constellations: Challenges for Space Law', in Froehlich, A. (ed.) *'Legal Aspects Around Satellite Constellations'*, Springer, 2019, p. 87.

[104] Ibid.

vehicle hardware after separation, and solid propellant slag.[105] Also included in this category are anticipated but not desired sources of debris, such as the release of paint flecks through the anticipated deterioration of the satellite. To a certain degree, existing ISO standards already regulate the production of anticipated debris, however introducing a hard law, enforceable requirement for all actors to abide by these and other applicable standards would be a worthwhile inclusion within the treaty.

Debris resulting from unplanned events refers to the risk of collisions. As noted, orbit congestion caused by large and mega SC, absent clarity in data sharing and detection/avoidance procedures, may lead to an increased risk of in orbit collision. Compounding this worry is that there is currently no universal coordinated system available to all actors through which orbiting objects can be tracked and potential collisions avoided. To fill this void, the treaty would establish a universal Space Traffic Management (STM) system. STM, as defined in the US's Space Policy Directive-3, National Space Traffic Management Policy (2018), is "the planning, coordination, and on-orbit synchronization of activities to enhance the safety, stability, and sustainability of operations in the space environment".[106] A universal STM, depending on the specifics of the system itself, could either facilitate, coordinate, or even perform the manoeuvring necessary to avoid potential collisions.

Currently, many operators have their own STM and collision detection and avoidance systems or procedures in place, usually synched to national databases, for example SpaceX's use of Department of Defense orbital debris data to inform its onboard automated detection and avoidance systems.[107] Establishing a universal STM to perform this function and making this system available to any and all operators reduces both inefficiency and room for error by removing the requirement of each actor to perform its own collision detection and avoidance monitoring.

Furthermore, universal STM could go beyond tracking orbiting objects and managing potential collisions. A universal STM would also provide an effective mechanism through which to establish common regulatory protocols relating to orbital management. In effect, by requiring the registration of all launched objects with the universal STM, policymakers could ensure that all objects abide by binding minimal standards embedded within the protocols of that STM. For example, this could allow policymakers to achieve alignment in maximum collision risk thresholds (past practice indicates that SpaceX accept a significantly higher risk of collision

---

[105] ISO, 'ISO 24113:2019 Space systems—Space debris mitigation requirements', www.iso.org/standard/72383.html.

[106] White House, 'Space Policy Directive-3, National Space Traffic Management Policy', 2018, www.whitehouse.gov/presidential-actions/space-policy-directive-3-national-space-traffic-management-policy/.

[107] Fernholz, T., 'SpaceX's new satellites will dodge collisions autonomously (and they'd better)', *Quartz*, 2019, https://qz.com/1627570/how-autonomous-are-spacexs-starlink-satellites/#:~:text=SPACEY-,SpaceX's%20new%20satellites%20will%20dodge,(and%20they'd%20better)&text=The%20first%20sixty%20Starlink%20satellites%20take%20flight%20on%20May%202023%2C%202019.&text=Today%2C%20Musk's%20space%20company%20said,and%20eventually%20more%20than%204%2C000.

than ESA's 1 in 10.000 threshold),[108] the codification of minimum requirements for on-board collision detection and avoidance systems (both manual and automated), or minimum standards for redundancies of propulsion systems.

Finally, and as the IADC research makes clear, the current rate of adherence to the Space Debris Mitigation Guidelines regarding end-of-life requirements is simply not high enough to ensure long-term sustainability of space. The treaty, through inclusion of a provision imposing a binding obligation on all actors to comply with both the UNOOSA guidelines and, potentially, the ISO standards, could solve this issue. Indeed, this would be the primary purpose of the treaty, to introduce a universal binding obligation on all actors to abide by the applicable regulatory frameworks.

Turning to the appropriate loci within which to pursue such a treaty, UNOOSA, and specifically COPUOS, is the ideal (and possibly the only) option. No other loci have the required mandate to facilitate the creation of an international treaty, nor the scope, expertise, and general competency to do so. The IADC would no doubt be a highly useful supporting locus, potentially as a forum through which to produce draft treaty text and secure initial international consensus on the scope and purpose of the treaty, however, the IADC has no mandate to produce the treaty itself. As noted, this treaty would be most effective if paired with revised, expanded and strengthened principles-based and rules-based frameworks. Here, the IADC, UNOOSA and ISO, as the organizations responsible for the current frameworks, are the obvious appropriate loci.

5 Loci for Future Space Regulation

The above are by no means an exhaustive list of the many distinct issues that constellations pose now or may pose in the future. Nonetheless, as representative case studies of the diversity of issues that may arise within the NewSpace context, they serve a valuable purpose within this article.

Through the construction of a conceptual framework of regulatory approaches, the authors demonstrated that a wide variety of options are available to policymakers in selecting an appropriate regulatory instrument for any given challenge. The distinct advantages and disadvantages of each of the potential approaches allows policymakers to pursue a bespoke regulatory strategy capable of achieving a range of desired outcomes, such as the codification of binding rules of behaviour or principles of outcomes. As the authors argued in section two, the range of options available offers policymakers the opportunity to combine multiple approaches in order to ensure a range of desired outcomes are met. There is no inherent reason why any given issue need only have a single regulatory solution, indeed multiple solutions may in some cases be the most effective strategy.

---

[108]ESA, 'Reentry and collision avoidance', www.esa.int/Safety_Security/Space_Debris/Reentry_and_collision_avoidance.

In section three, the authors employed this conceptual framework in examining the various bodies and institutions within industry, national and international levels able to serve as loci for regulatory solutions. Some loci were identified as being more capable of facilitating certain regulatory approaches than others, for example the ISO in producing rules-based frameworks, or COPUOS in the creation of international hard law. In general, the authors identified loci within the international level as being the most effective.

Finally, the authors applied the regulatory approaches framework and examination of loci to three issues concerning large and mega SC. Whereas the issue-approach-loci framing was employed primarily as an imperfect lens of analysis, with the relationships between issues, approaches and loci not being so linear in practice, this logic nonetheless yielded valuable insights into how a thorough analysis of the issue can inform regulatory approach and thereby selection of locus.

The scope of distinct issues posed by this one trend within NewSpace, itself a result of deeper and perhaps more transformational factors such as the democratization of knowledge and technology and increased ease and reduced costs associated with accessing space, indicates that policymakers must prepare themselves for both a high quantity and high variety of novel regulatory challenges in the space sector in the coming decades. Large and mega SC are only one space activity made ever more possible and beneficial for actors engaged in space. It should be expected that, as these more impactful factors increasingly transform the scope of human activity in space, many new challenges may arise.

The relative importance of international space policy will surely increase in tandem with the relative explosion of human activity in space, both commercial and not. Issues such as space mining rights, liability for space objects, violent non-state actor access to space and security of space-based assets are all set to become major challenges that must be addressed. The trends which have enabled the rise of NewSpace are not slowing down; in fact, they continue to accelerate. Identifying the best positioned loci for future regulatory efforts, adapting them, and expanding their mandate where appropriate, and, if necessary, creating new bodies capable of meeting the requirements of the situation is key to building a regulatory foundation capable of keeping pace with the ever-evolving scope of human activity in space.

Above all, this analysis has sought to emphasize and demonstrate that any single problem need not have only one solution. The diversity of regulatory approaches is a major advantage for policymakers. Dismissing one approach entirely in favour of another is unambitious and ultimately less effective than pursuing an issue from multiple fronts.

Jack B. P. Davies is a Research Affiliate at the Centre for Global Challenges (UGlobe), Utrecht University, and a Junior Fellow at the Human Security Centre. His research focuses primarily on issues relating to international security and public policy, including the role international organisations in global governance. Jack has previously written a variety of subjects ranging from the militarisation and weaponisation of space to the role of special operations forces in contemporary western security practice. He holds an MA in International Relations from the University of

Birmingham, having written his thesis on the issue of the use of armed drones under international law.

Jonathan Woodburn is a Young Graduate Trainee at the European Space Research and Technology Centre (ESTEC), working in the Software Product Assurance section for the European Space Agency. His research involves identifying issues between AI, machine learning and space quality assurance. He holds an M.Sc. in Computer Science from the University of Birmingham.

Potential Antitrust and Competition Challenges of Satellite Constellations

Alice Rivière

Abstract Satellite constellations have emerged as the poster child of the New Space business model. While the first satellite constellations were launched in the Nineties, the idea was revived lately and now a dozen of constellations and mega-constellations projects are planned for the current decade and thousands of satellites are planned to be launched in Outer Space. Those initiatives challenge the sustainability of the Outer Space environment and pose challenges for regulators at international and national level. One less obvious aspect of the heavily regulated satellite mega-constellation business model is the impact of antitrust regulations. Specifically, this article explores the potential frictions satellite mega-constellations might create in this area, focusing on U.S. antitrust laws and E.U. competition rules. After analyzing how antitrust and competition challenges arose and were handled in the traditional satellite industry, it will establish the elements that make satellite mega-constellations unique and how this might create challenges from an antitrust and competition law perspective.

1 Introduction

1.1 Emergence of a New Market Segment in the Satellite Industry Inducing New Regulations

"A next-generation space race is unfolding. We are seeing new commercial models, players, and technologies coming together to pioneer a wide range of cool satellite services. [..] However, this rush to develop new space opportunities requires new rules. Despite the revolutionary activity in our atmosphere, the regulatory frameworks we rely on to shape these efforts are dated." stated Jessica Rosenworcel,

---

The views expressed in this article should not be attributed to Airbus Defence and Space GmbH as the author wrote this chapter in her private capacity.

A. Rivière (✉)
Airbus Defence and Space, Munich, Germany

Commissioner of the U.S. Federal Communications Commission in the context of the spectrum licensing of SpaceX's Starlink mega-constellation in the U.S.[1]

Indeed, since the idea of commercial satellite constellations that emerged in the Nineties was revived by various companies a couple of years ago, commercial constellations have become the poster child of the predicted New Space model, crystallizing the announced transformations in the structure of the satellite market and setting new standards for the satellite industry.[2] As of early 2021, six companies have announced their ambitious projects to deploy large constellations of small satellites into low-earth orbit (LEO).[3] In addition, the European Commission announced its plan to launch its own satellite mega-constellation to operate its own "space-based communication system."[4] China is also said to plan a mega-constellation, but concrete plans have yet to be announced.[5] The potential for those constellations is huge and consequent improvements in space-based services are expected in three fields, namely, improving global coverage for internet-based communications, near real time measurements for Earth observation, space observation for continuous monitoring and surveillance.[6]

Those projects will have an enormous impact on the orbital landscape. According to the latest Euroconsult market forecast, a skyrocketing increase in the number of satellites launched is anticipated: an average of 1,250 satellites every year is expected to be launched in 2029, five times more than during the last decade.[7] Those

---

[1] Statement Of Commissioner Jessica Rosenworcel, 'Re: Space Exploration Holdings, LLC, Application for Approval for Orbital Deployment and Operating Authority for the SpaceX NGSO Satellite System, IBFS File No. SAT-LOA- 20161115-0018; Call Sign S2983; Application for Approval for Orbital Deployment and Operating Authority for the SpaceX NGSO Satellite System Supplement, SAT-LOA-20170726-00110, Call Sign S3018.

[2] A. Rivière, *The Rise of the LEO: Is There a Need to Create a Distinct Legal Regime for Constellations of Satellites?*, in: Annette Froehlich, Legal Aspects Around Satellite Constellations, Springer Studies in Space Policy, 40 (2019), p. 39-53.

[3] SpaceX was the first US-based entity authorized by the FCC to launch and operate Starlink, a massive broadband internet satellite constellation of 4425 broadband satellites orbiting approximately between 1100 and 1325 km in LEO; its concurrent OneWeb petitioned the FCC to access the US market with thousands of satellites authorized by the UK; Canadian satellite communication service provider Telesat also has a satellite internet project and plans deploy a constellation of 117 satellites in the LEO which was approved by the FCC in November 2017; Boeing and Leosat have also announced ambitious plans to put thousands of Internet-service satellites in the LEO. Finally, Amazon joined the constellation race in 2019 when its announced its plan to launch its large broadband internet constellation Kuiper.

[4] Jonathan O'Callaghan, *Europe Wants To Build Its Own Satellite Mega Constellation To Rival SpaceX's Starlink*, Forbes, 23 December 2020, www.forbes.com/sites/jonathanocallaghan/2020/12/23/europe-wants-to-build-its-own-satellite-mega-constellation-to-rival-spacexs-starlink/ (all websites cited in this publication were last accessed and verified on 24 January 2021).

[5] SpaceWatchGlobal, 4 October 2020, https://spacewatch.global/2020/10/china-pushes-ahead-with-giant-broadband-satellite-constellation.

[6] Giacomo Curzi, Dario Modenini, Paolo Tortora, *Large Constellations of Small Satellites: A Survey of Near Future Challenges and Missions*, Aerospace Review, 2020, 8.

[7] Euroconsult, Satellites to be build and launched by 2029, 'A complete analysis & forecast of satellite manufacturing and launched service', 2020 Edition.

projects are so advanced now that the changes they are bringing into the orbital landscape no longer go unnoticed in the night sky: the satellite train of the Starlink constellation is regularly observed by intrigued amateur spectators without telescope wondering what this chain of flashing white lights could be, guesses ranging from "rising stars in slow-motion" to an "airplane convoy".[8] Astronomers are concerned that those giant satellite clusters may create too many light streaks that will negatively interfere with astronomical observation, rendering astronomy impracticable in the very near future.[9] Satellite mega-constellation operators and astronomers are now working together to develop methods for assessing and reducing satellite brightness: recommendations both for operators and observatories have been published, such as implementing measures such as equipping satellites with sun-shielding visors or using image-cleaning software to strip out the light contamination.[10]

These constellation projects challenge the sustainability of the Outer Space environment and raise major issues of space debris, orbit congestion and saturation, and spectrum management.

In addition, satellite mega-constellations raise a host of legal, regulatory, and policy challenges that need to be addressed as those constellation projects are progressively being implemented and becoming reality.

In particular, this increasing number of satellites launched pose challenges for regulators at international and national level.

At international level, the International Telecommunication Union (ITU) decided during the WRC-19[11] on a milestone-based approach for the deployment of non-geostationary satellite systems including mega-constellations in LEO. For non-geosynchronous constellation operators to keep their full spectrum rights in the future, they will have to hit deployment milestones starting seven years after requesting the spectrum: after this period, constellation operators will need to launch 10% of their satellites in two years, 50% in five years and the rest of the constellation in seven years. If constellation ventures fail to launch enough satellites by the milestones, or within the total 14 years allotted, their spectrum rights will be limited proportionally to the number launched before time ran out.[12]

---

[8]La Depeche du Midi, 25 April 2020, www.ladepeche.fr/2020/04/25/limpressionnante-constellation-de-satellites-starlink-visible-dans-le-ciel-de-la-region-vendredi-soir,8862233.php.

[9]A. Witze, *How satellite 'megaconstellations' will photobomb astronomy images*, 26 August 2020, www.nature.com/articles/d41586-020-02480-5.

[10]Ibid. See for instance the National Science Foundation's NOIRLab and the AAS's Satellite Constellations 1 (SATCON1) report, 'Impact of Satellite Constellations on Optical Astronomy and Recommendations towards Mitigations', 21 July 2020, https://aas.org/sites/default/files/2020-08/SATCON1-Report.pdf.

[11]ITU regularly organizes World Radiocommunication Conferences ("WRC," known until 1992 as "WARC," World Administrative Radio Conferences), which are intended to examine the need for frequency allocation and attempt to apportion the spectrum in an equitable and forward-looking way while protecting the services already in place. The last WRC was held in Sharm el-Sheikh, Egypt in 2019.

[12]Press release, 'ITU World Radiocommunication Conference adopts new regulatory procedures for non-geostationary satellites Multiple satellite mega-constellations in low-Earth orbit to provide

At national level, the Federal Communications Commission (FCC) adopted in August 2019 a new streamlined licensing process for small satellites with shorter missions that require less intensive spectrum use and pose a lower orbital debris risk, allowing operators to secure spectrum rights in a much faster and cheaper way.[13]

As for any emerging growing market of promising size where only a handful of large companies and/or governments are active and have substantial market shares and that own important industry-standard-setting licenses, and technologies, so do grow in parallel concerns on the relations between competitors and issues of monopoly.[14] In addition, the heavy regulations imposed on operators in this area add another layer of complexity to those concerns since "[e]xperience has shown that regulated industries give rise to somewhat special types of monopoly and antitrust problems."[15]

One less obvious aspect of the heavily regulated satellite mega-constellation business model is the impact of antitrust regulations. Specifically, this article explores the potential frictions satellite mega-constellations might create in this area. The vast majority of those constellation projects are U.S. based, with the OneWeb network newly acquired by the UK Government and the Indian conglomerate BhartiUK/Indian.[16] For this reason, and because there is a de facto global antitrust regime based principally on the laws of the U.S. and the European Union,[17] for the purpose of this study, we will focus on U.S. antitrust and E.U. competition regulations.

After a brief overview of those two systems and notions (2), we will analyze how antitrust and competition challenges arose and were handled in the traditional satellite industry (3). Against this background, we will establish the elements that make satellite mega-constellations unique and how this might create challenges from an antitrust and competition law perspective (4).

---

extensive global telecommunications coverage', 20 November 2019, www.itu.int/en/mediacentre/Pages/2019-PR23.aspx.

[13] *Streamlining Licensing Procedures for Small Satellites*, IB Docket No. 18-86, Notice of Proposed Rulemaking, 33 FCC Rcd 4152 (2018) (*Small Satellite NPRM* or *NPRM*).

[14] Dariush Bahreini, Roerich Bansal, Gerd Finck, and Marjan Firouzgar, *Done deal? Why many large transactions fail to cross the finish line*, McKinsey, 5, August 2019, www.mckinsey.de/business-functions/strategy-and-corporate-finance/our-insights/done-deal-why-many-large-transactions-fail-to-cross-the-finish-line.

[15] 'Proceedings of the Conference on the Law of Space and of Satellite Communications.' NASA SP, 44. Washington, D.C., Scientific and Technical Information Division, National Aeronautics and Space Administration 82.

[16] *FCC transfers OneWeb's U.S. licenses and market access*, SpaceWatchGlobal, 14 November 2020, https://spacewatch.global/2020/10/fcc-transfer-oneweb-us-licenses-market-access/.

[17] "This antitrust regime is, in turn, reflected onto the rest of the world as other states adopt laws modeled on the U.S., the E.U. (increasingly more often the case) or a hybrid of U.S. and E.U. laws.", in: Eric Engle, 'The Globalization of Antitrust and Competition Law' (2012) 21 Currents: Int'l Trade LJ 3.

2 Overview of Antitrust and Competition Law

Antitrust or competition law is a field at the cross-section of law and economics. "With a proud origin in the American ideal of competition and free trade",[18] it is based on the assumption that, to operate as efficiently as possible, markets should ensure fair competition between companies to compete on level playing field. In particular, the market power of any particular firm should be limited not to distort competition. The idea underneath is that healthy competition serves the consumers' interests and gives them lower prices, higher-quality products and services, more choices, and promotes greater innovation. As explained by the Secretary-General of the Organisation for Economic Co-operation and Development (OECD) Angel Gurría: "Firms facing competitive rivals innovate more than monopolies [...] Competitive mechanisms can even help deliver on other strategic objectives, like environmental or health benefits. It all depends on good design. If companies are rewarded for producing the things we value, competition between them gives them the incentive to do so still better".[19]

Globally, antitrust law first emerged in the U.S. (2.1), before the European Union developed competition rules to protect its integrated single market (2.2.).

2.1 Antitrust: The US Perspective

U.S. antitrust laws finds in its origin in the 1890 Sherman Antitrust Act (the Sherman Act), the first U.S. federal statute to limit cartels and monopolies which still forms the legal basis for a significant portion of U.S. antitrust enforcement at a federal level.[20] The Sherman Act, the Federal Trade Commission Act of 1914, and the Clayton Act of the same year are the laws that set the corpus of antitrust regulation in the U.S.[21]

The Sherman Act focuses on the acquisition and obtention of monopoly power which is deemed to create in and of itself a threat for the market: its Sect. 2 even condemns the monopolies that remain unexercised.[22] On this basis, certain violations

---

[18] Richard C Stanley, 'Antitrust Law' (1990) 36 Loy L Rev 665.

[19] Remarks by Angel Gurría, OECD Secretary-General, at 'The Future Ain't What it Used to Be—20 Years of Competition Law and the Challenges Ahead', Reykjavík, Iceland, 17 September 2013, http://www.oecd.org/competition/20-years-of- competition-law-and-the-challenges-ahead.htm.

[20] ITU Report, 'Understanding patents, competition & standardization in an interconnected world', 39, 2014.

[21] See 'Competition Guidance, Guide to Antitrust Laws', the Federal Trade Commission: www.ftc.gov/tips-advice/competition-guidance/guide-antitrust-laws/antitrust-laws.

[22] Section 2 of the Sherman Act reads: "Every person who shall monopolize, or attempt to monopolize, or combine or conspire with any other person or persons, to monopolize any part of the trade or commerce among the several States, or with foreign nations, shall be deemed guilty of a felony, and, on conviction thereof, shall be punished by fine not exceeding one million dollars if a corporation, or, if any other person, one hundred thousand dollars, or by imprisonment not exceeding three years, or by both said punishments, in the discretion of the Court.", Sherman Antitrust Act, 15 U.S.C. § 2 (2006).

of antitrust law were soon declared to be per se illegal, such as splitting the market among producers by fixing production quotas or setting prices."[23] However, not all restraints on competition are deemed illegal and courts progressively developed a reasonable restraint test to determine whether the procompetitive effects of the limitation outweigh its anti-competitive effects.[24] In implementing this test, courts started to measure this reasonable restraint notion by examining its effect on consumers: if the effect of the restraint in question on the consumers is deemed beneficial, such as providing consumers with a greater number of goods of higher quality, and/or at lower prices, it is considered reasonable and the restraint is not censored.[25] To summarize, a measure is reasonable if it is efficient for the consumer and thus provides a net benefit to the market.[26]

The Federal Trade Commission Act bans "unfair methods of competition" and "unfair or deceptive acts or practices."[27]

Finally, the Clayton Act[28] focuses on preventing mergers and acquisitions that may "substantially lessen competition or tend to create a monopoly"[29] and preventing discriminatory prices, services and allowances in dealings between merchants.[30] The Act was notably amended by the Hart-Scott-Rodino Antitrust Improvements Act of 1976 requiring that companies file premerger notifications with the Federal Trade Commission and the Antitrust Division of the Department of Justice (DOJ) for certain acquisitions.[31]

As we shall now see, a different approach on competition distortions was developed in the European Union.

2.2 Competition Law: The E.U. Perspective

Within the European Union, the approach developed on those issues known as antitrust in the U.S. is called competition law and occurred in the context of the formation of a single integrated market after the Second World War. In this context, monopolies or cartels created in each different national market were deemed the

---

[23] See e.g. U.S. v. Topco Assocs., 405 U.S. 596, 607 (1972); N. Pac. Ry. Co. v. U.S., 356 U.S. 1, 5 (1958); Arizona v. Maricopa Cnty. Med. Soc'y., 457 U.S. 332, 334–45 (1982).

[24] Eric Engle, 'The Globalization of Antitrust and Competition Law' (2012) 21 Currents: Int'l Trade LJ 10.

[25] See e.g. Broad. Music, Inc. v. Columbia Broad. Sys., 441 U.S. 1 (1979).

[26] Eric Engle, 'The Globalization of Antitrust and Competition Law' (2012) 21 Currents: Int'l Trade LJ 10.

[27] 15 U.S.C. §§ 41–58.

[28] 15 U.S.C. §§ 12–27.

[29] 15 U.S.C. § 18—*Acquisition by one corporation of stock of another.*

[30] 15 U.S. Code § 14—*Sale, etc., on agreement not to use goods of competitor.*

[31] 15 U.S. Code § 18a—*Premerger notification and waiting period.*

principle obstacle to free competition in the single market.[32] As a result, the focus in the E.U. is not on the acquisition of monopoly power but rather "on the abuse of a dominant position on the market and also the use of antitrust in breaking up national markets in order to foster economic integration".[33] Thus, "abuse of dominant position" is the standard for determining violations of competition law in the E.U. and its Member States.[34] When determining the existence of anti-competitive behavior, the European Union put the emphasis on competitors, instead of efficiency and consumer welfare as U.S. courts and regulators do.

The economic rationale behind is, however, similar to the one in the U.S.: in the E.U. competition is considered a driver for economic growth in promoting productivity and innovation as well.[35] On this assumption, E.U. competition law considers that allowing collusive or monopolistic market behavior, such as cartels, to develop result in new firms being prevented from entering the market, which in the long term inhibits innovation.

In the E.U., competition rules are contained in various legal instruments. The basic provisions are contained in Articles 101 and 102 of the Treaty on the Functioning of the European Union (TFEU). Article 101 TFEU prohibits agreements or concerted practices between undertakings as well as decisions of associations of undertakings which restrict competition subject to some limited exceptions.[36]

---

[32] Eric Engle, 'The Globalization of Antitrust and Competition Law' (2012) 21 Currents: Int'l Trade LJ 10.

[33] Ibid. 13.

[34] Ibid.

[35] Directorate General For Internal Policies Policy Department, 'A: Economic And Scientific Policy, The Contribution Of Competition Policy To Growth And The E.U. 2020 Strategy Study', 10, 2013.

[36] Article 101 TFEU reads as follows: "1. The following shall be prohibited as incompatible with the internal market: all agreements between undertakings, decisions by associations of undertakings and concerted practices which may affect trade between Member States and which have as their object or effect the prevention, restriction or distortion of competition within the internal market, and in particular those which:

(a) directly or indirectly fix purchase or selling prices or any other trading conditions;

(b) limit or control production, markets, technical development, or investment;

(c) share markets or sources of supply;

(d) apply dissimilar conditions to equivalent transactions with other trading parties, thereby placing them at a competitive disadvantage;

(e) make the conclusion of contracts subject to acceptance by the other parties of supplementary obligations which, by their nature or according to commercial usage, have no connection with the subject of such contracts.

2. Any agreements or decisions prohibited pursuant to this Article shall be automatically void.

3. The provisions of paragraph 1 may, however, be declared inapplicable in the case of:

• any agreement or category of agreements between undertakings;

• any decision or category of decisions by associations of undertakings;

• any concerted practice or category of concerted practices,

which contributes to improving the production or distribution of goods or to promoting technical or economic progress, while allowing consumers a fair share of the resulting benefit, and which does not:

(a) impose on the undertakings concerned restrictions which are not indispensable to the attainment of these objectives;

Article 102 prohibits the abuse of a dominant position within the internal market.[37] The concept of 'abuse' is an "objective concept referring to the conduct of a dominant undertaking which is such as to influence the structure of a market where the degree of competition is already weakened precisely because of the presence of the undertaking concerned, and which, through recourse to methods different from those governing normal competition in products or services on the basis of the transactions of commercial operators, has the effect of hindering the maintenance of the degree of competition still existing in the market or the growth of that competition".[38]

A number of implementing Regulations have later been adopted, either by the Council or the European Commission. Importantly, merger control is governed by Council Regulation (EC) No 139/2004 of 20 January 2004 on the control of concentrations between undertakings (the EC Merger Regulation).

E.U. competition law thus focuses on three areas: prohibiting agreements, collaborations or practices restricting free trading or competition between businesses; prohibiting abusive conduct by a dominant market player; and monitoring market concentration and mergers.[39]

Admittedly, large aerospace companies involved in the satellite, and especially in the satellite communication business have been no strangers to antitrust and competition proceedings and problems.

(b) afford such undertakings the possibility of eliminating competition in respect of a substantial part of the products in question".

[37] Article 102 TFEU reads as follows: "Any abuse by one or more undertakings of a dominant position within the internal market or in a substantial part of it shall be prohibited as incompatible with the internal market in so far as it may affect trade between Member States. Such abuse may, in particular, consist in:

(a) directly or indirectly imposing unfair purchase or selling prices or other unfair trading conditions;

(b) limiting production, markets or technical development to the prejudice of consumers;

(c) applying dissimilar conditions to equivalent transactions with other trading parties, thereby placing them at a competitive disadvantage;

(d) making the conclusion of contracts subject to acceptance by the other parties of supplementary obligations which, by their nature or according to commercial usage, have no connection with the subject of such contracts.".

[38] Case 85/76 Hoffman-La Roche v Commission [1979] ECR 461, paragraph 91; Case C-62/86 AKZO v Commission [1991] ECR I-3359, paragraph 69; Case C-52/07 Kanal 5 and TV 4 [2008] ECR I-9275, paragraph 25; and Case C-52/09 TeliaSonera Sverige [2011] ECR I-527, paragraph 27.

[39] *See* ITU Report, 'Understanding patents, competition & standardization in an interconnected world', 39, 2014.

3 Past Antitrust and Competitions Challenges in the Traditional Satellite Market

As the satellite market was expanding, antitrust and competition concerns emerged in the satellite industry on both side of the Atlantic, but mainly crystallized around the issue of corporate merger clearances.

3.1 Evolution of Antitrust Aspects of in the U.S.

In the U.S. where regulators are attentive to any potential for monopolies, antitrust concerns arose as soon as regulating authorities were created in the satellite communication sector (3.1.1), but unfolded mainly when satellite companies undertook corporate mergers (3.1.2).

3.1.1 Initial Concerns

Antitrust concerns are intrinsically connected to the emergence of U.S. space ventures and emerged as soon as the U.S. put in March its space policies. Antitrust concerns were at the crux of the 1962 Communications Satellite Act signed by President Kennedy establishing the ownership, operation, and regulation of a commercial communications satellite system for the conduct of communications satellite operations: the publicly traded Communications Satellite Corporation (COMSAT).[40] The American Telephone and Telegraph Company (AT&T) was majority shareholder and generated fears of monopolistic behavior, while other shareholders were companies from different industries such as Hughes Aircraft Corporation, the Radio Corporation of America, ITT Corporation, General Electric, and Lockheed Martin.[41] A few weeks before the signature, on July 10, 1962, NASA had launched Telstar 1, an AT&T manufactured satellite that allowed the first live broadcast of television images between the U.S. and Europe.[42]

From the outset, Sect. 102(c) of the Communications Satellite Act enounced that competition was a paramount concern in the establishment of COMSAT: "It is the intent of Congress that all authorized users shall have nondiscriminatory access to the system; that maximum competition be maintained in the provision of equipment and services utilized by the system; that the corporation created under this Act be so organized and operated as to maintain and strengthen competition in the provision of communications services to the public; and that the activities of the corporation

---

[40] PUBLIC LAW 87-624-AUG. 31, 1962, 419.
[41] Whalen D.J. (2014) The Communications Satellite Act of 1962. In: The Rise and Fall of COMSAT 1.
[42] Ibid.

created under this Act and of the persons or companies participating in the ownership of the corporation shall be consistent with the Federal antitrust laws."[43]

It seems, however, that those antitrust concerns had little to do with the emerging satellite communication operations and their ground-breaking space technologies but focused first and foremost on the communication aspect induced by those novel satellite networks. The considerations enumerated as potential complicating factors from an antitrust perspective were the following:[44]

(1) Problems as to allocation of proprietary and other rights, as well as allocation of responsibilities between the Government and private industry, particularly in the light of increasing evidence of a strong need for extensive Government funds to support the research and development effort.

(2) The fact that initial costs and size of investment loom so large, and profits are so uncertain, that great initial encouragement may be given to monopoly.

(3) Problems arising out of affiliation between communications carriers and equipment manufacturers (most notably illustrated by the structure of AT&T).

(4) Competing claims and interests of different communications carriers-the largest and the not-so-large.

(5) Possible anticompetitive effects of creating a statutory joint venture in which competitors work together.

(6) Practical problems of assuring a proper degree of participation by "small business.

(7) Dangers of undue dominance accruing to those companies which have made a head start, which are represented on the board of the new corporation, which have access to all the information, and which in some instances may occupy what appears to be a most valuable "side track" position.

(8) Longer-range problems as to whether, when communications satellite systems have become established on a working and regular basis, there will be room for (or undue restraint upon) effective competition between the new corporation and the communication-carrier companies which own part of its stock and have a voice in its management.

Those concerns are not specific to nascent satellite communications satellite operations. Rather, they would apply to the creation of any public-owned joint venture consisting of competitors from different industries, with a majority shareholder with monopolistic inclinations. In addition, legal practitioners expressly insisted back then on the banality of satellite communications systems compared with telecommunications systems already in place: "in terms of service to the customer there is nothing that a satellite system can do that we cannot do in communications today, or that we will not be able to do with conventional facilities when the system comes into being."[45] Insisting on the fact that communications satellites would not "constitute

---

[43]PUBLIC LAW 87-624-AUG. 31, 1962, 419.

[44]Bennett Boslcey, 'Monopoly and Antitrust Aspects of Communications Satellite Operations, Proceedings of the Conference on the Law of Space and of Satellite Communications'. NASA SP, 44. Washington, D.C., Scientific and Technical Information Division, National Aeronautics and Space Administration 80.

[45]George V. Cook, 'Proceedings of the Conference on the Law of Space and of Satellite Communications'. NASA SP, 44. Washington, D.C., Scientific and Technical Information Division, National Aeronautics and Space Administration 100.

a communications system in themselves", but would integrate to existing terrestrial communication systems the same way "high-frequency radio and modern submarine cables" do, the conclusion was that they brought no novelty in terms of existing antitrust reasoning in the telecommunications realm. As one of the commentators declared: "[d]espite the glamour and excitement created by Telstar, technologically speaking communication via satellite is but another means of relaying, point-to-point, long-distance communications."[46] This fact that the satellite industry was treated no differently than other telecommunications from an antitrust perspective explains why the challenges in this field crystallized around the issue of corporate merger clearance. Indeed, the communications industry, primarily in the areas of telecommunications, cable television, and studio entertainment, is presently one of the most active merger and acquisition markets in the U.S.[47]

3.1.2 Antitrust Clearances in the Traditional Satellite Market

The telecommunication sector was no stranger to the antitrust protection emerging in the U.S. at the beginning of the twentieth century. Some of the first and most prominent decisions by courts censuring monopolistic corporate behavior concern telecommunications company. The example of AT&T can be cited here. In 1949, a complaint was filed against AT&T and Western Electric contending that both companies had "had been engaged in a continuing conspiracy to restrain and monopolize the manufacture, distribution, and sale of telephones, telephone apparatus, and telephone equipment in violation of Sects. 1 and 2 of the Sherman Act"[48]

The Communications Satellite Act of 1934 which grants the FCC the authority to regulate the use of the entire electromagnetic spectrum,[49] also bestow the institution with the duty to enforce certain antitrust principles to ensure effective competition in the procurement of equipment and services by COMSAT.[50] Also, the Communications Satellite Act empowers and obligates the FCC to determine whether the transfer of control of a station license in the context of a merger or acquisition proceeding

---

[46] Ibid.

[47] Robert B. Friedrich, *Regulatory and Antitrust Implications of Emerging Competition in Local Access Telecommunications: How Congress and the FCC Can Encourage Competition and Technological Progress in Telecommunications*, 80 Cornell L. Rev. 647 (1995).

[48] *See* Report of the Antitrust Subcommittee of House Judiciary Committee on Consent Decree Program of the Department of Justice, ch. II: The A.T.&T. Consent Decree, pp. 29 et seq. (86th Cong., 1st Sess. 1959). The suit was settled in 1956 by the Government's giving up the demand for divestiture and also giving up the alternative demand for the limitation of the role of Western as an exclusive supplier of AT&T, and settling for provisions opening up some patents for general licensing.

[49] Proceedings of the Conference on the Law of Space and of Satellite Communications. NASA SP, 44. Washington, D.C., Scientific and Technical Information Division, National Aeronautics and Space Administration 98.

[50] Carl H. Fulda, 'Proceedings of the Conference on the Law of Space and of Satellite Communications'. NASA SP, 44. Washington, D.C., Scientific and Technical Information Division, National Aeronautics and Space Administration 91.

serves the "public interest, convenience, and necessity.[51] This analysis comes in addition to the merger review by the Antitrust Division of the DOJ pursuant to the Clayton Act. The FCC's competitive analysis is broader and goes further since it is not limited to checking whether the proposed merger will reduce existing competition, but also considers its effect on "the market power of dominant firms in the relevant communications markets and the transaction's effect on future competition"."[52] An example of this FCC mandate in the satellite industry was the 2007 proposed merger between Sirius Satellite Radio Inc. and XM Satellite Radio Holdings Inc. in the realm of satellite radio that was approved both by the DOJ and the FCC.[53]

Another prominent example of a satellite corporate merger is the case of COMSAT itself: on September 20, 1998, Lockheed Martin Corporation announced that it would merge with COMSAT which became effective after receiving FCC approval on August 3, 2000.[54] In its review of the proposed merger, the FCC enounced the test to established the competitive effects of the proposed transaction occurring in the realm of telecommunications service: after declaring that it had to consider "both the relevant product and geographic markets", the Commission determined that "the relevant product markets can include both service to U.S. domestic telecommunications markets and service between the U.S. and international telecommunications markets."[55] It then cleared the proposed merger after noting that they had found "no evidence of potential harm to competition resulting from the proposed transaction, and in view of these stated benefits, we find that based on the claimed benefits, the proposed transaction would be in the public interest.[56]

As the commercial satellite industry grows and matures, consolidation of global actors to create a game-changing global satellite operators has been brought up and discussed for decades, not ever to avail.[57] Antitrust clearances on the context

---

[51] Section 310(d) of the Communications Act reads: "No construction permit or station license, or any rights thereunder, shall be transferred, assigned, or disposed of in any manner, voluntarily or involuntarily, directly or indirectly, or by transfer of control of any corporation holding such permit or license, to any person except upon application to the Commission and upon finding by the Commission that the public interest, convenience, and necessity will be served thereby. 47 U.S.C. § 310(d).

[52] DirecTV-Liberty Media Merger, 23 F.C.C.R. at 3278.

[53] Press Release, 'Federal Communications Commission, Commission Approves Transaction Between Sirius Satellite Radio Holdings Inc. and XM Satellite Radio Holdings Inc. Subject to Conditions' (July 28,2008).

[54] Lockheed Martin, COMSAT Combination Completed, 3 August 2000, Defense Aerospace: http://www.defense-aerospace.com/article-view/release/2790/lockheed-completes-comsat-acquisition-%28aug.-4%29.html.

[55] In the Matter of Lockheed Martin Corporation, COMSAT Government Systems LLC, And COMSAT Corporation, COMSAT Corporation and Its Subsidiaries, Applications for Transfer of Licensees of Various Satellite, Earth Station Private Land Mobile Radio and Experimental Licenses214 Authorizations, and Holders of International Section, Control of File No. SAT-T/C-20000323-00078, File No. SAT-STA-20000323-00073.

[56] Ibid.

[57] J. Hill, *It Was Supposed to be a Perfect Match: An Autopsy of Three Failed Satellite Industry Takeovers*, http://interactive.satellitetoday.com/via/January-2021/it-was-supposed-to-be-a-perfect-match-an-autopsy-of-three-failed-satellite-industry-takeovers/?utm_campaign=coschedule&utm_source=twitter&utm_medium=Via_Satellite.

of corporate merger routinely occurred in the satellite sector, it has even been observed that,"communications sector mergers are also twice as likely to face antitrust challenges as mergers in other sectors."[58]

3.2 Evolution of Competition Aspects in the E.U.

In the E.U., competitions concerns emerged later in the satellite industry than they did in the U.S. This finds its explanation in the history of the European Union (called back then European Communities) that first accelerated its unification and expansion in the 1980s and 1990s.[59]

3.2.1 Competition Concerns During the Liberalization of the Satellite E.U. Market

Recognizing that satellite communications were an integral and essential part of the "new global information highway", and observing that satellite communications were moving world-wide into a "new phase of development", the need to liberalize the satellite market was progressively recognized throughout the 1990s.[60] Competition in the satellite market was considered as restricted or distorted by government regulation allowing special or exclusive rights to particular bodies.[61]

In October 13, 1994, the Commission adopted the Satellite Liberalisation Directive requiring the abolition of all exclusive rights granted for the provision of satellite services and equipment, and the abolition of all special rights to provide any telecommunications service covered by the Directive.[62] The aim was to stimulate the greater use of satellite communications within the E.U. by lifting the existing restrictions and fostering competition in the satellite market where Members States where the main actors.[63] The fact that satellite services was regulated to only allow particular

---

[58] Dariush Bahreini, Roerich Bansal, Gerd Finck, and Marjan Firouzgar, *Done deal? Why many large transactions fail to cross the finish line*, McKinsey, August 5 2019: https://www.mckinsey.de/business-functions/strategy-and-corporate-finance/our-insights/done-deal-why-many-large-transactions-fail-to-cross-the-finish-line.

[59] *See* The history of the European Union, https://europa.eu/european-union/about-eu/history_en.

[60] The Council Resolution of 22 July 1993 committed the E.U.-Member States to the opening of public voice telephony to competition by 1 January 1998. In November 1994 the Council agreed to the Commission's proposal to extend the 1998 liberalization to include the lifting of restrictions on infrastructure for the provision of telecoms services, *see* Dr H. Ungerer, 'Regulatory Directions For Satellite Communications In Europe, Intelsat Summit London', 2 March 1995, https://ec.europa.eu/competition/speeches/text/sp1995_012_en.html.

[61] Dr H. Ungerer, 'Regulatory Directions For Satellite Communications In Europe, Intelsat Summit London', 2 March 1995, https://ec.europa.eu/competition/speeches/text/sp1995_012_en.html.

[62] Directive 94/46/EC.

[63] Dr H. Ungerer, 'Regulatory Directions For Satellite Communications In Europe, Intelsat Summit London', 2 March 1995, https://ec.europa.eu/competition/speeches/text/sp1995_012_en.html.

bodies was deemed a distortion of competition that was no longer necessary in the interests of public service.[64]

At the same time, and to further enhance these efforts of liberalization and to ensure fair competition, the need to increase and harmonize copyright protection in the satellite broadcasting domain was recognized. The E.U. passed a directive governing the application of copyright and related rights to satellite and cable television to "ensure that any difference in the level of protection within the common market will not create distortions of competition".[65]

Alongside these competition developments in the services side of the satellite market, in the E.U. as well competitions concerns crystallized at the stage of corporate mergers.

3.2.2 Competition Aspects of Merger Clearance in the Traditional Satellite Industry

Under E.U. competition law, mergers and acquisition must comply with the provisions of Article 4 of the EC Merger Regulation.[66] The EC Merger Regulation defines two categories of merger, i.e. "concentrations" targeted: those arising from a merger between previously independent undertakings, and those arising from an acquisition of control.[67] In the context of a competition clearance, it is "necessary to establish whether or not concentrations with a Community dimension are compatible with the common market in terms of the need to maintain and develop effective competition in the common market".[68]

Several so-called concentrations occurred in the traditional satellite industry that gave the European Commission the opportunity to conduct competitive analysis in this sector. We can cite three cases, two of which concern Luxemburg company Astra.

Notably, in 1992, the Commission censored existing joint-venture agreements between SES and British Telecommunications (BT) whereby the two parties cooperated in the joint provision of a television distribution service by satellite and deemed that it constituted an infringement of Article 85 (1) of the Treaty establishing the European Economic Community (ECC Treaty), the former Article 101 TFEU[69]

---

[64]Ibid.

[65]Council Directive 93/83/EEC of 27 September 1993 on the coordination of certain rules concerning copyright and rights related to copyright applicable to satellite broadcasting and cable retransmission.

[66]"Article 4 Prior notification of concentrations and pre-notification referral at the request of the notifying parties 1. Concentrations with a Community dimension defined in this Regulation shall be notified to the Commission prior to their implementation and following the conclusion of the agreement, the announcement of the public bid, or the acquisition of a controlling interest.".

[67]Article 3(1) of the Merger Regulation.

[68]Recital 23 of the Merger Regulation.

[69]93/50/EEC: Commission Decision of 23 December 1992 relating to a proceeding pursuant to Article 85 of the EEC Treaty (IV/32.745—Astra).

The Commission did a competitive analysis the proposed joint-venture aiming at offering "operators of UK-originated TV programmes a packaged service consisting of a BT uplink in the UK and transponder space on SES's satellite."[70] The Commission first declared that SES and BT were indeed "direct competitors in the European market for the provision of space segment capacity for the transmission of television channels".[71] On this basis, it concluded that this partnership between uplinker BT and satellite provider SES for a bundled service was a distortion on competition and a "restrictive horizontal agreement"[72] because "customers who wished to transmit their programmes via the Astra IA satellite were not given the choice of concluding separate contracts for, on the one hand, uplink services and, on the other hand, the lease of transponder capacity".[73] Had they been two different contracts, customers would have had room for negotiating different prices. As a result, the Commission ordered that BT and SES inform programme providers who signed contracts with BT for international TV distribution services via the Astra IA satellite to either renegotiate their terms or terminate.[74]

Besides this broadcasting case which led to the analyze of the space segment of the proposed joint-venture, other decisions concerning mergers in the satellite communication and broadcasting services industry did not prompt the Commission to refuse the proposed mergers.

In the satellite-based broadcasting sector, another decision concerning Astra was examined by the Commission. More than a decade after it censored the joint-venture between SES and BT, the Commission examined a joint-venture proposal between SES (that had become Astra S.A. in the meantime) and Eutelsat S.A. which would be active in the provision of infrastructure for both broadcasting content to mobile devices and two-way voice and data communication services to mobile devices.[75] After noting that the transaction would constitute a "concentration" within the meaning of Article 3(1)(b) of the Merger Regulation,[76] the Commission undertook a competitive analysis of the primary business of the-venture: implementing an infrastructure for broadcasting content to mobile devices. In particular, it noted that the joint-venture proposal would be "active in the market for broadcasting content to mobile devices, whereas the JV's parents have activities in broadcasting content to fixed devices which belong to a distinct market so that there would be no directly competing activities between the two."[77] It concluded that the "proposed operation

---

[70] Ibid., para. 6.

[71] Ibid., para. 12.

[72] Ibid., para. 15.

[73] Ibid., para. 33.

[74] Ibid., para. 33.

[75] Case No Comp/M.4477, Ses Astra/Eutelsat/JV Notification of 20 June 2007 pursuant to Article 4 of Council Regulation No 139/2004.

[76] Ibid., para. 8.

[77] Ibid., para. 19.

would not give raise to serious competition concerns in the market for the provision of infrastructure for broadcasting content to mobile devices."[78]

The Commission then analyzed the other aspect of the proposed joint-venture: the satellite- based infrastructure for two-way mobile voice and data communication services for mobile (handheld) devices for maritime, aeronautical and land-based applications. It concluded that because there were enough 'strong competitors" in this market segment "the proposed transaction is unlikely to give rise to serious doubts in the market for the provision of infrastructure for two-way voice and data communication services.[79] Finally, the Commission allowed the joint-venture declare it compatible with the common market.[80]

Another competition clearance by the Commission in the satellite communications sector was the acquisition of Telenor Satellite Services (TSS) by Apax Partners France in 2006.[81] The Commission undertook a competitive analysis in each of the area of use, i.e. aeronautical, land and maritime in each applicable geographic market and concluded that "the notified operation and to declare it compatible with the common market".[82]

The Commission undertook competitive analysis of the space segments of the proposed deals and showed that it considered and gave enough weight to the satellite services model to acknowledge its peculiarities in comparison to other telecommunications means.

As a matter of fact, the peculiarity of mergers and acquisitions in the satellite sector is now increasingly recognized by legal instruments in Europe. The update of the Luxemburg Space Law contains a special provision targeting acquisitions and change of control in space activity operating companies which must be notified to the ministry in charge of space activities.[83]

---

[78]Ibid., para. 28.

[79]Ibid., para. 31.

[80]Ibid., para. 40.

[81]Telenor Satellite Services to be acquired by Apax Partners, 26 October 2006: www.telenor.com/media/press-release/2006/telenor-satellite-services-to-be-acquired-by-apax-partners.

[82]Case No COMP/M.4709 – Apax Partners/ Telenor Satellite Services, Notification of 13 July 2007 pursuant to Article 4 of Council Regulation No 139/2004.

[83]Loi du 15 décembre 2020 portant sur les activités spatiales. Article 13(1) reads as follow: "Toute personne physique ou morale qui a pris la décision d'acquérir ou d'augmenter, directement ou indirectement, une participation qualifiée dans un opérateur, avec pour conséquence que la proportion de parts de capital ou de droits de vote détenue atteindrait ou dépasserait les seuils de 20 pour cent, de 30 pour cent ou de 50 pour cent ou que cet opérateur deviendrait sa filiale, informe à l'avance et par écrit le ministre de son intention".

4 Antitrust and Competition Challenges for Upcoming Satellite Mega-Constellations?

As we have seen, the antitrust and competition challenges faced by the traditional satellite market were for the most part similar to those faced by other telecommunications means. The question is whether satellite constellations will bring novel issues that will in turn raise new antitrust and compliance challenges.

One important challenge faced by satellite constellations is their financial costs and tendency to go bankrupt.[84] This characteristic has yet to translate into peculiar antitrust or competitions problems. Two competition clearances of mergers occurring as a consequence of bankruptcy occurring in the U.S., namely in the Iridium case at the dawn of the 2000s and recently in October 2020 in the OneWeb case.[85] No specific consideration were given to the network and float of satellite side of the constellation projects in neither case. For instance, in the Iridium case, the FCC applied the test previously developed for satellite telecommunications service providers in its competitive assessment: "For satellite telecommunications service providers, the Commission has determined that the relevant product markets include domestic and international telecommunications markets. In cases involving such service providers, the Commission considers whether the proposed transaction will lessen or enhance competition in the provision of communications services in, to or from the United States."[86]

Another prominent aspect of the satellite constellation model is the scale of the new mega-constellations proposals and astonishing number of satellites to launch. Until now, the assumption was that space was so large that satellite operations would not interfere with each other, but this may not be true in the near future as we enter into an era of scarce orbital slot opportunities. Satellite constellations pose major challenges to the sustainability of Outer Space.

With such a number of active elements in orbit, their management will become a fundamental point of interest. In particular, intensive intersatellite communication and coordination between actors will become key points. To avoid creating overcrowded radiofrequency spectrum and physical interferences with adjacent radiofrequency signals and maximize traffic capacity of the communication infrastructure, mega-constellations will need to solve the challenges of fleet management, communication, and space tracking.[87] Technical solutions like spectrum sharing are being

---

[84] By the beginning of 2000, OrbComm, Iridium, and Globalstar had all gone bankrupt and filed for Chapter 11 protection in the U.S. while Teledesic suspended its activities in 2002.

[85] Application Granted For Assignment And Transfer Of Control By Worldvu Satellites Limited, Debtor-In-Possession, IB Docket No. 20-290, IBFS File Nos. SES-ASG-20200818-00891, SAT-MPL-20200818-00099, and SAT-MPL-20200831-00101.

[86] In re Applications of Space Station System Licensee, Inc., Assignor, And Iridium Constellation LLC, File No. SAT-ASG-20010319-00025 for Consent to Assignment of License), Pursuant to Sect. 310(d) of the) Communications Act, 8 February 2002, para. 33.

[87] Giacomo Curzi, Dario Modenini, Paolo Tortora, *Large Constellations of Small Satellites: A Survey of Near Future Challenges and Missions*, Aerospace Review, 2020, 6.

pushed forward where temporarily unused spectrum could be reallocated for a more efficient use.[88] Most importantly, physical coordination between systems will be required to avoid space debris creation as emphasize by the FCC in all its licensing decisions authorizing the launch of mega-constellation, such as in the case of SpaceX: "While we are concerned about the risk of collisions between the space stations of NGSO systems operating at similar orbital altitudes, we think that these concerns are best addressed in the first instance through inter-operator coordination."[89]

Coordination has always played an essential role in ITU's licensing system which always relied on its member states to exercise goodwill and mutual assistance in their space endeavors.[90] Some companies planning satellite constellations such as OneWeb and Boeing went as far as concluding orbiting settlements to regulate the distance at which their respective constellations would orbit.[91] Such private agreements to share spectrum and orbital positions holdings to carve markets between competitors could well be considered anti-competitive agreements within the meaning of Article 101 TFEU. More generally, divulging or exchanging business-sensitive information for constellation management and coordination purposes with other actors active in the same orbits concerned could potentially constitute a form of concerted business practices prohibited by competition law. This conduct could also qualify as an intentional horizontal limit to competition between competitors and, thus as an infringement of the Federal Trade Commission Act.

Another practice relating to satellite constellations that could create competition and antitrust issues is the so-called spectrum warehousing practice whereby satellite companies delay the launch of their satellites or buy so-called "paper filings" to hoard the spectrum. This could constitute a potential abuse of a dominant position within the internal market within the meaning of Article 102 TFEU. If a satellite constellation operating company is in a dominant market position and uses spectrum warehousing technique to keep or enhance its dominant position, courts of the European union would likely condemn this practice. Notably, the Court of Justice of the European Union (CJEU) concluded in a medical patent case that a medical company's attempt to mislead the patent offices amounted to an abuse of a dominant position and that the deregistration of the marketing authorizations with the intention of preventing generic market entry was inconsistent with European competition law because it excluded from the market competing manufacturers of generic products.[92]

---

[88] Höyhtyä, M.; Mämmelä, A.; Chen, X.; Hulkkonen, A.; Janhunen, J.; Dunat, J.-C.; Gardey, J., *Database-Assisted*.

Spectrum Sharing in Satellite Communications: A Survey, 2017, 5, 25322–25341.

[89] FCC 18-38, 2018, IBFS File No. SAT-LOA-20161115-00118—Application for Approval for Orbital Deployment and Operating Authority for the SpaceX NGSO Satellite System, III. 11, www.fcc.gov/document/fcc-authorizes-spacex-provide-broadband-satellite-servicesaccessed.

[90] J. Zoller, Improving the international satellite regulatory framework, 2011.

[91] *See* P. de Selding, OneWeb, Boeing settle constellation orbit issue; SpaceX questions OneWeb ownership, Space Intel Report blog entry, 25 April 2017, www.spaceintelreport.com/oneweb-boeing-settle-constellation-orbit-issue-spacex-questions-oneweb-ownership/.

[92] AstraZeneca AB & AstraZeneca plc v European Commission, Court of Justice of the European Union, 6 December 2012, C-457/10 P, ECLI:E.U.:C:2012:770.

Finally, a last aspect of satellite constellations that might create antitrust or competitions issues is standardization. As constellation projects progressively mature in an environment where coordination is key, standardization will inevitably develop. Standardization is a form of collaboration among competing market players to develop and codify best practice in the interests of encouraging widespread adoption which inherently bears potential anti-competitive effects.[93] In general, standardization is a good thing for competition because it allows interoperability which has beneficial effects for the consumer in a market as highly regulated as the telecommunication one. Anticompetitive effects could be created if the standard-setting organization does not ensure fair, reasonable, and non-discriminatory licensing (FRAND).[94]

5 Conclusion

Satellite constellations are at the intersection of many regulations and rules and raise many questions, in particular concerning the sustainability of Outer Space, as the projects are progressively being manufactured and launched. As it appears, satellite constellations can raise antitrust and competition challenges which operators must bear in mind even though regulators have not yet dealt with this problematic. However, the context of orbital scarcity that creates these potential challenges in the realm of antitrust and competition law could well act as their legitimate justification.

Indeed, in the E.U., an interesting movement started in the realm of competition law in the past years that could have groundbreaking consequences in the field. Recent developments emerged around the idea that "it was time to reconsider the relationship between sustainability and antitrust to ensure companies can cooperate on what it is "good for the planet"."[95]

Sustainability of Outer Space is the major challenge of the New Space era and the necessity to cooperate on what is good for Outer Space might pardon anti-competitive practices.

Alice Rivière is admitted as an attorney-at-law in the State of New York and as *avocat à la cour* in Paris. She works in the Legal and Compliance Department of Airbus Defence and Space in Munich, Germany. She previously practiced law in the field of international dispute settlement in Switzerland.

---

[93] ITU, Understanding patents, competition & standardization in an interconnected world, 32, 2014.

[94] FRAND standards are voluntary licensing commitment that are often requested from the owner of an intellectual property right (usually a patent) that is, or may become, essential to practice a technical standard, see Layne-Farrar, Anne; Padilla, A. Jorge; Schmalensee, Richard, *Pricing Patents for Licensing in Standard-Setting Organizations: Making Sense of FRAND Commitments,. Antitrust Law Journal.* 74 671 (2007).

[95] Grant Murray, Antitrust and sustainability: globally warming up to be a hot topic?, Kluwer Competition Law Blog, 18 October 2019, http://competitionlawblog.kluwercompetitionlaw.com/2019/10/18/antitrust-and-sustainability-globally-warming-up-to-be-a-hot-topic/.

She holds an LL.M. in International Law from the University of Miami, a Master's degree in International Economic Law from Université Paris I Panthéon-Sorbonne/Columbia Law School/Sciences Po Paris, a Maîtrise en droit from Université Paris I Panthéon-Sorbonne, and an LL.M./Magister legum from Universität zu Köln.

The Designation of Satellite Constellations as Critical Space Infrastructure

John Tziouras

Abstract Near-Earth space is becoming increasingly privatized and industrialized, with many consequences for science and humanity. In particular, the number of satellites in low-Earth orbit is predicted to grow dramatically from a couple of thousands at present to many tens of thousands in the near future due to the launch of satellite constellations planned by public and private entities. Large satellite constellations may create new legal and security challenges. Efforts to reduce satellite constellations risks, especially those emanating from natural hazards or intentional attacks, requires an adaptation in both critical infrastructures and space policies. Therefore, in order to realise the full potential of investments in space, critical systems need to be adequately protected and the space environment properly managed. In terms of security of critical space infrastructures past approaches have been largely oriented towards protection. However, there is evidence to suggest a transition promoting the resilience of critical space infrastructures, including those with space segments such as satellites.

1 Introduction

Satellites have become the backbone of a wide variety of space applications related to communications, navigation, meteorology and remote sensing. Many governmental and military activities, as well as societal functions, heavily rely on space-based assets. Any temporary or permanent disruption of these functions, whether intentional or not, may have severe consequences ranging from reduced economic productivity to a loss of human life.[1]

---

[1] Marcus Matthias Keupp, 'The Security of Critical Infrastructures: Introduction and Overview' in Marcus Matthias Keupp (ed), *The Security of Critical Infrastructures: Risk, Resilience and Defense* (Springer Nature Switzerland AG 2020) 1.

J. Tziouras (✉)
Faculty of Law (LL.M.), Aristotle University of Thessaloniki, Thessaloniki, Greece
e-mail: tziouras@law.auth.gr

In this context, humanity's increasing dependence on satellites places them firmly in the field of critical infrastructures (CIs).[2] Since satellites have become vital components of national and international security, this inclusion is even more warranted. Deficient or inadequate critical infrastructure protection (CIP) may affect national and international security, the economy and social well-being.

Although satellites are listed merely as key assets under the protection umbrella of various national CIP policies, recognition of their significance is gaining momentum. In a space environment that is not only congested and contested but also subject to heavy investment, the deployment of large satellite constellations offers new challenges.

This article aims to introduce satellite mega-constellations into the concept of CIP and define them as critical space infrastructures (CSIs).[3] As comprehensive multi-sectoral national policies to support the resilience or protection of CSIs have begun to emerge, several legal and security issues related to critical space infrastructure protection (CSIP) and large satellite constellations must be examined.

2 Addressing Threats Related to the Emergence of Large Satellite Constellations

As satellites continue to be placed into orbit, space assets have become increasingly vulnerable to loss due to threats of natural (i.e., random) or human (i.e., intentional or malevolent) origin. Apart from anti-satellite weapons (ASATs), two unique threats—space debris and space weather—are known to have a direct impact on satellites and terrestrial systems that depend on satellites, some of which are critical.[4] The current section outlines the characteristics of space debris and the impact of space weather in low Earth orbit (LEO)—that is, an orbit close enough to Earth for convenient transportation, communications and observation of Earth. This is an area that the International Space Station currently orbits and in which proposed large satellite constellations will be located in the future.[5] As the analysis of intentional threats such as ASATs and other forms of interference with satellite systems (i.e., cyberattacks) fundamentally differs from the analysis of unintentional threats (since probabilistic risk analysis is inappropriate when risk originates from an intelligent adversary), this study focuses exclusively on random occurrences of disruptive events.

---

[2] Markus Hesse and Marcus Hornung, 'Space as Critical Infrastructure' in Kai-Uwe Schrogl et al. (eds), *Handbook of Space Security: Policies, Applications and Programs* (Springer Science + Business Media New York 2015) 188.

[3] Alexandru Georgescu et al. (eds), *Critical Space Infrastructures: Risk, Resilience and Complexity* (Springer Nature Switzerland AG 2019) 37.

[4] Ibid.

[5] Darcy Elburn, 'Low-Earth Orbit Economy' (*NASA*, 15 Jan 2021) <www.nasa.gov/leo-economy/faqs> (All websites cited in this publication were last accessed and verified on 15 January 2021).

2.1 Orbital Debris in Low-Earth Orbit

Space is not an empty vacuum. It contains both natural debris (i.e., meteorites) and human-made space debris (i.e., non-operational spacecraft, abandoned launch vehicle stages and satellite fragmentations). The use and exploration of space over the past 60 years has cluttered the area around Earth with an enormous quantity of human-made space debris, many of which can remain in orbit for years or even centuries.[6]

There is no internationally agreed legal definition of space debris, and the term is not mentioned in any of the space treaties.[7] With high orbital velocities that sometimes exceed 10 km/s in LEO, satellites are increasingly at risk of collision with orbital debris or other satellites. As the number of satellites or orbital population increases, the possibility of disruptive collisions cannot be discounted.

Whilst satellites are routinely replaced at the end of their lifetime, orbits cannot be. Under the assumption that orbits are as important as the assets (i.e., satellites) that occupy them, any loss of an orbit's usefulness due to the accumulation of space debris would render the assets of this finite resource useless.[8]

In a December 2020 filing, an American communication company called Viasat once again requested that the U.S. Federal Communications Commission (FCC) either conduct an environmental assessment or make a more rigorous environmental impact statement about SpaceX's Starlink before approving the company's request to modify its existing license for the system in order to operate more satellites in lower orbits.[9] Part of the petition addressed orbital debris. Despite SpaceX's citation of statistics that claimed a low failure rate for its Starlink satellites, Viasat have been a strident critic of their reliability in FCC filings and have expressed concerns that satellites that fail in orbit could add to the growing population of debris in LEO.

The LEO environment is becoming increasingly crowded. It has been estimated to contain more than 500,000 uncontrolled orbiting objects larger that 1 cm in diameter.[10] Several recent projection studies have indicated that the debris population in some regions of LEO will become unstable, because the collision risk for a mixed population (either satellites or debris) is not only the sum of the collision risk of each individual space object but must also account for interactions between satellite

---

[6] Joe Pelton et al. 'Space Safety' in Kai-Uwe Schrogl et al. (eds), *Handbook of Space Security: Policies, Applications and Programs* (2nd edn Springer Nature Switzerland AG 2020) 277.

[7] Annette Froehlich, 'The Right to (Anticipatory) Self-Defence in Outer Space to Reduce Space Debris, in: Annette Froehlich (ed), *Space Security and Legal Aspects of Active Debris Removal* (Springer Nature Switzerland AG 2019) 73.

[8] Adrian V. Gheorghe and Daniel E. Yuchnovicz, 'The Space Infrastructure Vulnerability Cadastre: Orbital Debris Critical Loads' (2015) 6 International Journal of Disaster Risk Science https://link.springer.com/article/10.1007/s13753-015-0073-2.

[9] Viasat Inc Petition to Deny or Defer (2020) https://ecfsapi.fcc.gov/file/12221423419154/Viasat%20Ex%20Parte%20Letter%20(12-21-2020).pdf.

[10] Space Environment Statistics, (*ESA*, 8 January 2021) https://sdup.esoc.esa.int/discosweb/statistics/.

and debris in the same population.[11] In a 2017 study on the effects of the OneWeb constellation, Radtke et al. found that, at an 800 km altitude, each satellite had a 69.35% chance of colliding with an object of 3 cm or larger during its lifetime.[12] They estimated that a single collision increased the debris flux for other satellites in the same constellation by a factor of nine. Furthermore, it has been posited that a non-catastrophic collision between a space object and a piece of debris at 10 km/s would result in an ejected mass equivalent to 115 times the mass of the smaller object.[13]

With regard to large satellite constellations consisting of hundreds to thousands of 100–300 kg class spacecraft in LEO, a 2018 National Aeronautics and Space Administration (NASA) Orbital Debris Program Office study showed that mass distribution in LEO would be dominated by spacecraft and upper stages (i.e., rocket bodies), which would dramatically change the landscape of satellite operations in that region.[14]

At the same time, the LEO economy is driven by a growing market for global and real-time connectivity.[15] Several commercial companies have already announced or begun the deployment of large constellation of small satellites in LEO. New LEO satellite concepts offer faster communications and often provide higher bandwidth per user than geosynchronous (GEO) satellites—even higher than cable, copper and pre-5G fixed wireless.[16] Organizations such as UBS and Morgan Stanley have projected that the space economy will be worth a total of U.S. $ 1.1 billion by 2040,[17] with space-based internet and related services in LEO accounting for at least 50% of this figure. Thus, the LEO economy is increasingly becoming a multi-billion-dollar question.[18]

---

[11] R. Lucken and D. Giolito, 'Collision Risk Prediction for Constellation' (2019) 161 Acta Astronautica www.sciencedirect.com/science/article/abs/pii/S0094576518321878?via%3Dihub.

[12] Jonas Radtke et al., 'Interactions of the Space Debris Environment with Mega-Constellations: Using the Example of the OneWeb Constellation' (2017) 131 Acta Astronautica www.sciencedirect.com/science/article/abs/pii/S009457651630515X.

[13] Tanja Masson-Zwaan and Mahulena Hofmann, *Introduction to Space Law* (4th edn, Kluwer Law International BV, The Netherlands) 109.

[14] J. C. Liou et al., 'NASA ODPO's Large Constellation Study' (2018) 22 (3) Orbital Debris Quarterly News www.orbitaldebris.jsc.nasa.gov/quarterly-news/pdfs/odqnv22i3.pdf.

[15] Alice Riviere, 'The Rise of the LEO: Is There a Need to Create a Distinct Legal Regime for Constellations of Satellites?' in: Annette Froehlich (ed), *Legal Aspects Around Satellite Constellations* (Springer Nature Switzerland AG 2019) 42–43.

[16] Chris Daehnick et al., 'Large LEO Satellite Constellations: Will it be Different this Time?' (*McKinsey & Company*, 4 May 2020) www.mckinsey.com/industries/aerospace-and-defense/our-insights/large-leo-satellite-constellations-will-it-be-different-this-time#.

[17] 'Space: Investing in the Final Frontier' (*Morgan Stanley*, 24 July 2020). www.morganstanley.com/ideas/investing-in-space.

[18] Keith W. Crane et al., 'Measuring the Space Economy: Estimating the Value of Economic Activities in and for Space' (Institute for Defense Analysis 2020) www.ida.org/-/media/feature/publications/m/me/measuring-the-space-economy-estimating-the-value-of-economic-activities-in-and-for-space/d-10814.ashx.

Consequently, any major collision in LEO would not only lead to a loss of function or the complete destruction of a satellite in a large constellation, but it also could trigger a chain reaction (the so-called 'Kessler Syndrome') that could render the entire orbit unusable.[19] According to a recent study, if only one primary collision occurred at an altitude of 800 km, the probability of a collision involving a constellation satellite would become greater than 2% by 2035, which would significantly jeopardize the satellite constellation as a whole.[20]

Whilst the continued operation of a satellite could be taken for granted today, any failure—temporary or otherwise—could have real consequences, especially in the case of a large-scale catastrophic event. Moreover, such consequences could affect the operability of other systems due to their interdependence with other infrastructures; these include space or terrestrial systems, some of which are responsible not only for society's well-being but also for physical existence of humans on Earth.[21]

2.2 Space Weather in Low-Earth Orbit

Similarly, space weather has a global footprint and can simultaneously affect multiple infrastructures. Electrical grids, telecommunications and wireless networks, spacecraft and satellite navigation services could all be seriously affected. An event of such magnitude could overwhelm more than one nation's response capacity. Historical evidence has shown that many infrastructures in space and on the ground are vulnerable to the effects of space weather.[22]

In September 2019, the European Space Agency (ESA) announced for the first time that it had raised the orbit of its 'Aeolus' satellite in order to avoid a collision with a SpaceX Starlink satellite.[23] The orbital 'conjunction' took place at approximately 318 km above sea level in LEO, with an estimated minimum distance of less than 1 km at the time of the nearest approach. The primary cause of this conjunction was not a miscalculation of orbital mechanics but space weather.[24]

---

[19]Donald J. Kessler and Burton G. Cour-Palais, 'Collision Frequency of Artificial Satellites: The Creation of a Debris Belt' (1978) 83 (A6) Journal of Geophysical Research https://agupubs.onlinelibrary.wiley.com/doi/abs/10.1029/JA083iA06p02637.

[20]R. Lucken and D. Giolito, 'Collision Risk Prediction for Constellation' (2019) 161 Acta Astronautica www.sciencedirect.com/science/article/abs/pii/S0094576518321878?via%3Dihub.

[21]Adrian V. Gheorghe et al. (eds), *Critical Infrastructures, Key Resources, Key Assets: Risk Vulnerability, Resilience, Fragility, and Perception Governance* (Springer International Publishing AG 2018) 28.

[22]Elisabeth Krausmann et al., 'Space Weather & Critical Infrastructures: Findings and Outlook' (EUR 28237 November 2016) https://publications.jrc.ec.europa.eu/repository/bitstream/JRC104231/space_weather_cover+report_final.pdf.

[23]ESA Spacecraft Dodges Large Constellation (*ESA*, 3 September 2019) www.esa.int/Safety_Security/ESA_spacecraft_dodges_large_constellation.

[24]T. E. Berger et al., 'Flying Through Uncertainty' (2020) 18(1) Space Weather https://agupubs.onlinelibrary.wiley.com/doi/10.1029/2019SW002373.

Space weather is the main source of uncertainty in the position of all objects below around 1000 km in LEO. The main impact, inter alia, is strong variation in the neutral density of the atmosphere as it responds to radiative inputs from the Sun, as well as global-scale electrical currents generated during geomagnetic storms.[25]

According to a study by the U.K. Royal Academy of Engineering, up to 10% of satellites could experience temporary outages that last hours to days as a result of an extreme space weather event.[26] In addition, significant cumulative radiation doses are expected to cause rapid ageing of satellites. Consequently, after an extreme event, satellite owners and operators will need to carefully evaluate the need for replacement satellites to be launched earlier than planned in order to mitigate the risk of premature failures. This assumption introduces new variables for the lifetime of large satellite constellations and post-mission disposal.[27]

SpaceX is already executing autonomous manoeuvres within the Starlink constellation, but no serious progress has been made to date to efficiently prepare for a future in which thousands of satellites could simultaneously move to new orbits. Without significant improvements in space situational awareness (SSA), space traffic management (STM) and space weather forecasting (as well as their coordination), there is a real risk of an increase in cascading collisions that could render not only satellites but even LEO unusable for decades or possibly even centuries.[28]

Although the danger posed by orbital debris and space weather has been characterized by some observers as a long-term environmental problem that affects space assets, others have perceived threats to national security interests, especially a state's ability to ensure consistent satellite support to national military and intelligence organizations.

Specifically, a U.S. Congressional Research Service report from July 2014 mentioned that, as orbital debris in some orbits (i.e., LEO) had reached a critical point beyond which the situation was defined as unstable, 'this instability in the space environment present a threat to U.S. national interest in space'.[29] Furthermore, according to the 2018 Presidential Space Policy Directive-3, the 'United States has effectively reaped the benefits of operating in space to enhance their national security'. Regarding the significance of space activities, 'the United States considers the

---

[25] Ibid.

[26] Royal Academy of Engineering, *Extreme Space Weather: Impacts on Engineered Systems and Infrastructure* (Royal Academy of Engineering 2013) www.raeng.org.uk/publications/reports/space-weather-full-report.

[27] J. C. Liou et al., 'NASA ODPO's Large Constellation Study' (2018) 22(3) Orbital Debris Quarterly News www.orbitaldebris.jsc.nasa.gov/quarterly-news/pdfs/odqnv22i3.pdf.

[28] T. E. Berger et al., 'Flying Through Uncertainty' (2020) 18(1) Space Weather https://agupubs.onlinelibrary.wiley.com/doi/10.1029/2019SW002373.

[29] U. S. Congressional Research Service, *Threats to U.S. National Security Interests in Space: Orbital Debris Mitigation and Removal* (Report No 7-5700, 2014) www.hsdl.org/?view&did=748309.

continued unfettered access to and freedom to operate in space of vital interest to advance the security of the Nation'.[30]

In light of these considerations, the following section discusses the extension of the concept of national security by including other spheres than the military in its meaning—namely, certain types of threats that frame a new security paradigm under CIP and particularly CSIP.

3 National Security, Critical Infrastructures Protection and Space Security

Traditionally, 'security' refers to a set of measures taken by a state or a coalition of states to protect and promote their national interests. In this context, states would consider themselves to be secure if they could survive and win the war. However, the emergence of new threats renders this approach less practical, as military forces are re-conceptualized from an instrument of war to a tool of deterrence.

With the inclusion of non-traditional sectors such as technology, the environment and even outer space in the security domain, the security agenda has broadened, gradually leading to the paradigm shift needed to address new threats. Along with the extension of the concept of national security through the inclusion of sectors other than military ones, the attention of experts has turned to entities that contribute to the well-being of citizens and addresses their vital needs. Institutions that are responsible for food, water, energy, transportation and communications—the so-called CIs—have become visible, but their vulnerability and the difficulty of protecting them against asymmetric threats have also become evident.[31] Since the presence of powerful military forces no longer represents a guarantee for social well-being, states soon realized that they would not be able to defend all of their CIs, no matter how powerful they were.

Whilst all infrastructures are, or eventually become, visible in the sense of capturing public interest, an infrastructure qualifies as critical only when people feel that its failure would effectively represent a clear and present danger to their well-being.[32] The damage or destruction of CIs—which are essential for maintaining vital societal functions—through natural disasters or intentional attacks may have negative consequences for the security of a state and the well-being of its citizens.

---

[30] Space Policy Directive-3, 83 F.R. 28969 (2018) www.federalregister.gov/documents/2018/06/21/2018-13521/national-space-traffic-management-policy.

[31] Adriana Alexandru et al., 'National Security and Critical Infrastructure' (2019) 25(1) International Conference Knowledge-Based Organization https://content.sciendo.com/view/journals/kbo/25/1/article-p8.xml?product=sciendo.

[32] Adrian V. Gheorghe et al. (eds), *Critical Infrastructures, Key Resources, Key Assets: Risk Vulnerability, Resilience, Fragility, and Perception Governance* (Springer International Publishing AG 2018) 27.

To reduce the vulnerability of CIs, states such as the United States and organizations such as the European Union (E.U.) have launched CIP policies. Such initiatives aim to strengthen the security and resilience of vital CIs.

However, CIs do not exist in isolation. There is increasing interdependence between infrastructure systems, as the functions of many facilities and services depend on others.[33] Given the nature of these systems, the analysis of CIs requires a system of systems perspective.[34] Consequently, it is assumed that the disruption of any of these infrastructures could jeopardize the continued operation of entire sectors under a cascading failure approach. Thus, the protection of such systems is the main aim of CIP.

Space assets, particularly satellites, have become deeply embedded in the functioning of vital societal services such as security and economic well-being. The increasing dependence on certain space systems places them firmly in the domain of CIs, leading some scholars to include them in the existing CIP framework as CSIs and under CSIP. There have also been proposals to include large satellite constellations in the concept of CSIs.[35]

4 The Designation of Satellite Constellations as Critical Space Infrastructure

Space systems, including satellites, rockets, space probes, and space and terrestrial stations, offer vital services that range from cheap, constant and instantaneous communications with worldwide coverage to navigation, remote sensing and other applications (e.g., financial transactions, industrial control systems such as supervisory control and data acquisition, management of energy grids and air traffic control). All of these systems are vulnerable to risks and threats that can have a real impact on societies.[36] As these systems represent a system of systems that is essential for the maintenance of vital governmental or societal functions, their destruction or disruption could have a significant, large-scale impact on any given nation.

Arguably, under this perspective, a satellite could represent a CSI, not only because it is located in space but also due to its connection with other vital functions, which

---

[33] Alexandru Georgescu et al. (eds), *Critical Space Infrastructures: Risk, Resilience and Complexity* (Springer Nature Switzerland AG 2019) 4.

[34] Ibid 4.

[35] Adrian V. Gheorghe et al. (eds), *Critical Infrastructures, Key Resources, Key Assets: Risk Vulnerability, Resilience, Fragility, and Perception Governance* (Springer International Publishing AG 2018); Alexandru Georgescu et al. (eds), *Critical Space Infrastructures: Risk, Resilience and Complexity* (Springer Nature Switzerland AG 2019); Kai-Uwe Schrogl et al. (eds), *Handbook of Space Security: Policies, Applications and Programs* (2nd edn, Springer Nature Switzerland AG 2020).

[36] Adrian V. Gheorghe et al. (eds), *Critical Infrastructures, Key Resources, Key Assets: Risk Vulnerability, Resilience, Fragility, and Perception Governance* (Springer International Publishing AG 2018) 27.

cross many dependent sectors of CIs.[37] Consequently, a satellite loss could affect the operability of other systems that share an interdependence with CSIs.

The geographic and economic realities of satellites systems, especially in the case of large satellite constellations, call for a different approach to protection efforts and resilient strategies based on international cooperation and collective action. The next section introduces the need to designate large satellite constellations as components of CSIs.

4.1 Critical Space Infrastructures. A System of Systems Approach

The criticality of an infrastructure is assessed in terms of the effects of its impact in a given time span, even a very short one.[38] The assessment of a CI can be based on criteria such as physical or functional. In addition, CI analysis must account for the multi-dimensional nature of infrastructures by considering both their engineering properties (i.e., physics-based properties that shape and constrain the operation of an infrastructure) and behavioural properties (i.e., relational properties that emerge from business processes, decision points, human interventions and participating information).[39] All of the above characteristics contribute to making CI analysis a complex matter. Thus, a system of systems approach to CSI is needed.

A system of systems approach is appropriate for understanding large-scale and highly complex phenomena comprised of highly interdependent participating systems, which themselves may be large-scale and highly complex. Such a phenomenon can be described as a system of systems when the system's behaviour is reflected in the emergent, synergistic behaviours of participating systems.[40] Critical space infrastructures possess these characteristics, as each system encompasses highly complex technologies and information.

Defense Support Program (DSP) has been the stalwart of missile warning since the 1970s.[41] For more than 40 years, U.S. satellites were used to detect missile launches under the DSP—so-called 'early warning satellites'. This system has been replaced by the more advanced Space-Based Infrared System (SBIRS), an integrated system of systems consisting of ground components, main system satellites and

---

[37] Ibid., Alexandru Georgescu et al. (eds), *Critical Space Infrastructures: Risk, Resilience and Complexity* (Springer Nature Switzerland AG 2019).

[38] Adriana Alexandru et al., 'National Security and Critical Infrastructure' (2019) 25 (1) International Conference Knowledge-Based Organization https://content.sciendo.com/view/journals/kbo/25/1/article-p8.xml?product=sciendo.

[39] William J. Tolone et al. (eds), 'Enabling System of Systems Analysis of Critical Infrastructure Behaviors' in Setola R., Geretshuber S. (eds) *Critical Information Infrastructure Security*. CRITIS 2008 Lecture Notes in Computer Science 5508 (Springer 2009).

[40] Ibid

[41] Edward P. Chatters and Bryan Eberhardt, 'Missile Warning Systems' in AU-18 Space Primer (Air University Press 2009) www.jstor.org/stable/resrep13939.24?seq=2#metadata_info_tab_contents.

additional satellites which provide broader coverage for more accurate and timely event reporting.[42] A malfunction in one of the ten main satellites that carry SBIRS or Space Tracking and Surveillance System payloads could jeopardize the entire early warning infrastructure, resulting in national or even international security concerns.

Similarly, as a system of systems, space-based telecommunication satellites interact with the information communication technology CI sector, providing immense capacity for digital data, storage, processing and more. This makes their combined interaction via ground segments interdependent.[43] Consequently, even a temporary loss of satellite services could lead to a CI cascading effect, resulting in a significant impact on the economy and society. Therefore, under the system of systems approach, space infrastructures depend on extensive interconnections with each other.

The criticality of satellites as space infrastructures is even greater, because they are located in one of the most hostile environments. As previously mentioned, the threats of space debris and space weather can have a direct impact on satellite systems and, consequently, to terrestrial services that depend on satellites. Therefore, an integrated model approach is required to provide security and reliability assessment that considers various kind of threats and failures.

As Georgescu et al. have argued, CSIs constitute a limited pool of significant assets that concentrate valuable services for a multitude of users.[44] Thus, CSIs interdependencies are fundamental considerations when assessing the resilience of key assets and resources of space infrastructures.

In U.S. legislation, a key resource is defined as 'a publicly or privately controlled resource essential to the minimal operations of the economy and government'.[45] Similarly, key assets are characterized as every possible individual target whose attack could result not only in large-scale human casualties and property destruction in the worst-case scenarios but also in profound to national prestige, morale and confidence.[46]

These key assets and resources are both worth protecting under CIP. Therefore, in the context of CSIs, emerging CSIP encompasses the protection of all space assets or resources. Thus, satellites could comprise space infrastructures, the information

---

[42] Pat Norris, 'Satellite Programs in the USA' in Kai-Uwe Schrogl et al. (eds), *Handbook of Space Security: Policies, Applications and Programs* (2nd edn, Springer Nature Switzerland AG 2020) 1157.

[43] Harald Opitz, 'Default Risk of Satellite Based Critical Infrastructure' (Security Research Conference, Berlin 2016) www.researchgate.net/publication/308900858_DEFAULT_RISK_OF_SATELLITE_BASED_CRITICAL_INFRASTRUCTURE.

[44] Alexandru Georgescu et al. (eds), *Critical Space Infrastructures: Risk, Resilience and Complexity* (Springer Nature Switzerland AG 2019) 22.

[45] U.S. Homeland Security Act of 2002, § 2, www.dhs.gov/sites/default/files/publications/hr_5005_enr.pdf.

[46] The White House, *National Strategy for the Physical Protection of Critical Infrastructures and Key Assets* (2009) www.dhs.gov/xlibrary/assets/Physical_Strategy.pdf.

that they handle could be viewed as key resources and their orbit could be seen as key assets.[47]

Given the abovementioned threats, numerous government reports, laws and executive orders have addressed CIP policies and strategies by using the terms 'key assets' and 'key resources' to define fundamental elements of a CI that are worth protecting. The following section examines the evolution of the concept of CSIP according to the existing CIP framework, which forms an integral part of national sustainability strategies. In addition, the section identifies requirements for large satellite constellations to be designated as CSIs that are worth protecting.

4.2 The Designation of Satellites as Critical Space Infrastructure

Although CIs were first identified in a 1983 U.S. Congressional Budget Office report,[48] it was only in the 1990s that a commission on CIP was established at the national level under Presidential Executive Order 13010.[49] The commission's report highlighted the interdependence between infrastructures and information and communication systems, as well as the need to view this through a national security lens.[50] Furthermore, satellites were included as a critical part of information and communication infrastructures.

The September 11 attacks demonstrated the vulnerability of CIs. In response, the United States strengthened its efforts to regulate this type of risk.[51] Since then, federal policy on CIP has been established through laws, presidential directives and national strategies.[52]

More specifically, a new executive order was signed in October 2001 under President George Bush's administration. The executive order established a new national

---

[47] Adrian V. Gheorghe et al. (eds), *Critical Infrastructures, Key Resources, Key Assets: Risk Vulnerability, Resilience, Fragility, and Perception Governance* (Springer International Publishing AG 2018) 28.

[48] Congress of the United States, Congressional Budget Office, *Public Works Infrastructure: Policy Considerations for the 1980s* (CBO Study, 1983) www.cbo.gov/sites/default/files/98th-congress-1983-1984/reports/doc20-entire.pdf.

[49] U.S. Presidential Exec. Order No. 13010, 3 C.F.R. 37345 (1996) www.hsdl.org/?view&did=1613.

[50] The White House, *Critical Foundations: Protecting America's Infrastructures* (Report, 1997) https://fas.org/sgp/library/pccip.pdf.

[51] Markus Hesse and Marcus Hornung, 'Space as Critical Infrastructure' in Kai-Uwe Schrogl et al. (eds), *Handbook of Space Security: Policies, Applications and Programs* (Springer Science + Business Media New York 2015) 191.

[52] U.S. Congressional Research Services, *Critical Infrastructure Protections: The 9/11 Commission Report and Congressional Responses* (Report No RL 32531, 2005) https://fas.org/sgp/crs/homesec/RL32531.pdf.

CIP strategy. Later, another executive order established the Office of Homeland Security, whose main mission was to 'develop a comprehensive national strategy to secure the U.S. from terrorist threats or attacks'.[53]

One year later, the first National Strategy for Homeland Security was released, and the Homeland Security Act of 2002 was signed into law.[54] The act established the Department of Homeland Security (DHS), which was assigned the task of coordinating national CIP efforts.[55] Specifically, the Homeland Security Act gave DHS the responsibility to conduct comprehensive assessments on the vulnerability of the United States' key resources and CIs and to develop a comprehensive national plan to secure them.

In both the first National Strategy for Homeland Security and the Homeland Security Act, information and telecommunications were identified as CI sectors. The Information and Communication Technology (ICT) sector was considered to be particularly important, because it connects and helps to control many other infrastructure sectors.

In August 2002, a report issued by the U.S. General Accounting Office highlighted the importance of satellites in national governmental and economic activities. In the report, it was mentioned that 'in light of the nation's growing reliance on commercial satellites to meet military, civil, and private sector requirements, omitting satellites from the nation's approach to protecting critical infrastructures leaves an important aspect of our nation's infrastructures without focused attention'.[56] This report was the first of its kind to identify the significance of satellites for the ICT sector and to introduce them into the CIP agenda.

In February 2003, the White House released the National Strategy for the Physical Protection of Critical Infrastructures and Key Assets, which specifically mentioned the key assets of the telecommunications sector. According to the strategy, the telecommunications sector faced significant challenges in the new threat environment with regard to protecting its vast and dispersed critical assets. Whilst the government and other CI industries heavily relied on the Public Switched Telecommunications Networks (PSTN), part of which included satellites, the sector's protection initiatives were particularly important.[57] Similarly, Internet infrastructure, which consists of Internet Service Providers (ISPs) that provide end users with Internet access, was also characterized as an important component of the telecommunications infrastructure.[58]

---

[53] U.S. Presidential Exec. Order No 13228, 3 C.F.R. 51812 (2001) www.hsdl.org/?abstract&did=1619.

[54] U.S. Office of Homeland Security, *National Strategy for Homeland Security* (2002) www.dhs.gov/sites/default/files/publications/nat-strat-hls-2002.pdf.

[55] Homeland Security Act of 2002, § 101, www.dhs.gov/sites/default/files/publications/hr_5005_enr.pdf.

[56] U.S. General Accounting Office, *Critical Infrastructure Protection: Commercial Satellite Security Should Be More Fully Addressed* (Report No 02-781, 2002) www.govinfo.gov/content/pkg/GAOREPORTS-GAO-02-781/html/GAOREPORTS-GAO-02-781.htm.

[57] The White House, *National Strategy for the Physical Protection of Critical Infrastructures and Key Assets* (2009) www.dhs.gov/xlibrary/assets/Physical_Strategy.pdf.

[58] Ibid.

In subsequent years, the DHS developed several National Infrastructures Protection Plans (NIPPs). In the 2009 version, it identified 16 CI sectors. For the first time, the crosscutting nature of space applications was recognized in the plan, and space-based and terrestrial positioning, navigation and timing services were viewed as components of multiple CIs and key resources sectors.

This recognition was also visible in the Cybersecurity and Infrastructure Security Agency Act of 2018, in which satellites were explicitly included in the key resources and CIs of the United States.[59]

In April 2019, a new industry group was established in the United States to share intelligence on cyber threats to space-based assets such as satellite communications. The Space Information Sharing and Analysis Center (Space-ISAC) is the newest of 22 ISACs and lobbied to the Donald Trump administration to designate commercial space systems as CIs. Information Sharing and Analysis Centers (ISACs) are non-profit organizations that provide a central resource for gathering data on cyber threats (mainly to CIs) and two-way information sharing between the private and public sectors. At the time of writing, the form in which this designation may continue under the new Joe Biden administration was unclear.

Although the space sector has not been officially recognized by the DHS as one of the 16 CI sectors covered under NIPPs in the United States, satellites have been designated as key resources and CIs that are worthy of protection in national legislation efforts.

The E.U.'s history with CIP features some similar events as the United States, which have highlighted the need for a European strategy that focuses on the protection of CIs. The September 11, Madrid and London attacks also help to explain why terrorism has been identified as the most important source of threats.

From 2004 onwards, the E.U. became increasingly involved in homeland security and CIP. In 2005, a Green Paper was released on European Programme for CIP; one year later, the first Communication from the European Commission on CIP was drafted.[60] In December 2009, the Council of the E.U. adopted Directive 2008/114/EC on the identification and designation of European CIs and the assessment of the need to improve their protection.[61]

With regard to CIs in space, the European Commission (EC) first broached the topic at the E.U. level in its 2011 Communication about a space strategy for the E.U.[62] According to the communication, space infrastructures should be considered

---

[59] U.S. Cybersecurity and Infrastructure Security Agency Act of 2018, www.congress.gov/bill/115th-congress/house-bill/3359.

[60] Markus Hesse and Marcus Hornung, 'Space as Critical Infrastructure' in Kai-Uwe Schrogl et al. (eds), *Handbook of Space Security: Policies, Applications and Programs* (Springer Science + Business Media New York 2015) 197.

[61] Alessandro Lazari, *European Critical Infrastructure Protection* (Springer International Publishing Switzerland 2014) 48, Council Directive 2008/114/EC of 8 December 2008 on the identification and designation of European critical infrastructures and the assessment of the need to improve their protection [2008] OJ L 345/75.

[62] Commission, 'Towards a Space Strategy for the European Union that Benefits its Citizens' (Communication) COM (2011) 152 final.

CIs and protected; furthermore, such protection was a major issue for the E.U. and outstripped the individual interests of satellite owners.

Other European states, such as the United Kingdom and France, have officially recognized space and satellites as CIs in their CIP policies. In particular, the United Kingdom's 2015 National Space Policy mentions that, as space has become increasingly important to modern Britain, 'space assets are rightly recognized as part of [its] critical national infrastructure'.[63] Similarly, according to the 2019 French Space Defence Strategy and previous presidential decrees,[64] France designated space as one of the 12 CI sectors worth protecting, along with research.[65]

According to Hesse and Hornung, space applications should be seen and treated as CIs.[66] Other scholars, such as Georgescu et al., have argued for space systems to be treated as a new CI category.[67] Since space systems have already been acknowledged as components of existing CI domains, the concept of CSIs is a natural outgrowth that is expected to enhance security efforts.

Since a growing number of both private and governmental space actors plan to develop large satellite constellations, their designation as CSIs should be examined.

4.3 Satellite Constellations as Critical Space Infrastructures

Interest in large satellite constellations is not confined to the United States and commercial space actors. The E.U., Russia and China are all pursuing their own proliferated constellation projects – the so-called 'mega-constellations'.

In December 2020, the EC selected a consortium of European satellite manufactures, operators, service providers, telecom operators and launch service providers to study the design, development and launch of a European-owned space-based communication system. The study will examine how the space-based system could enhance and connect to current and future CIs, strengthening the E.U.'s capability to access the cloud and provide digital services in an uninterrupted and secure way. This is essential for building confidence in the digital economy and ensuring European strategic autonomy and resilience.

---

[63] H.M. Government, *National Space Policy* (Policy Paper, 2015) www.gov.uk/government/publications/national-space-policy.

[64] France 'Arrêté du 2 juin 2006 fixant la liste des secteurs d'activités d'importance vitale et désignant les ministres coordonnateurs desdits secteurs' JORF n. 0129 du 04/06/2006, www.legifrance.gouv.fr/download/pdf?id=5R0NnW587uliImx9srlKXu-nam6aCtsgM2LdqywZyGE=.

[65] The French Ministry for the Armed Forces, *Space Defence Strategy* (2019).

[66] Markus Hesse and Marcus Hornung, 'Space as Critical Infrastructure' in Kai-Uwe Schrogl et al. (eds), *Handbook of Space Security: Policies, Applications and Programs* (Springer Science + Business Media New York 2015) 198.

[67] Alexandru Georgescu et al. (eds), *Critical Space Infrastructures: Risk, Resilience and Complexity* (Springer Nature Switzerland AG 2019).

As previously mentioned, there is a growing demand for LEO orbits, which significantly increases security complexity. The addition of hundreds of thousands of proliferated constellation satellites could increase congestion, stress the existing space situational awareness and space traffic management capabilities of states and lead to a more dangerous debris environment. Unless satellite constellations become more reliable, they could pose a long-term threat to the ability of powerful space actors to freely operate in space.[68] On these accounts, there is a need for approaches that can address large satellite constellations as CSIs.

The identification and designation of national CIs is mandated in various national and international regulations, as previously mentioned. This designation can be implemented in four stages.[69]

The first stage concerns the identification of sectors and subsectors that are considered to be important for national interests. As previously discussed, the space sector fulfils all of the requirements needed to be seen and treated as a distinct type of CI: a CSI. Thus, satellites constitute components of a CSI, which has already been recognized by some states.

The second stage entails the identification of critical services. All communication services, including space-based and terrestrial ICT applications, have already been recognized as critical and designated as official CI sectors in several national CIP strategies. In this context, internet services that are provided by satellites and vital to governmental and societal activities are considered to be critical services.

The third stage concerns the designation of CIs. For each service, the critical assets or components that comprise the CI must be identified and designated. Under the CSI approach, information or services provided by satellites could be recognized as key assets of CSIs.

In the fourth and final stage, procedures for protection and security are implemented for each CI as part of a comprehensive CIP policy. Critical infrastructure protection is a concept that relates to preparedness, readiness and response to serious incidents that involve space infrastructures, key resources and assets.

Under Georgescu et al.'s CSI taxonomy, communication satellites usually include certain subsystems, such as communication payloads (i.e., antennas), engines used to bring satellites to the desired orbit, tracking and stabilization subsystems and command and control subsystems.[70] All of these components are vulnerable. As a result, any disruption of CSI activity can have a direct impact on other CIs, not only under the system of systems approach but also through their connection relationship.

The architecture of mega-constellations appears to have additional vulnerabilities compared to current-generation satellite constellations. In particular, constellation

---

[68] Matthew A. Hallex and Travis S. Cottom, 'Proliferated Commercial Satellite Constellations: Implications for National Security' (2020) 97 JFQ https://ndupress.ndu.edu/Portals/68/Documents/jfq/jfq-97/jfq-97_20-29_Hallex-Cottom.pdf?ver=2020-03-31-130614-940.

[69] Nikolaos Petrakos and Panayiotis Kotzanikolaou, 'Methodologies and Strategies for Critical Infrastructure Protection' in Dimitris Gritzalis et al. (eds), *Critical Infrastructure Security and Resilience: Theories, Methods, Tools and Technologies* (Springer Nature Switzerland AG 2019) 19.

[70] Alexandru Georgescu et al. (eds), *Critical Space Infrastructures: Risk, Resilience and Complexity* (Springer Nature Switzerland AG 2019) 45.

satellites in LEO must avoid collisions with both operational satellites and non-operational space debris. The advent of mega-constellations will require a massive increase in collision avoidance manoeuvres, which could swell from a current average of three per day to approximately eight per hour. In October 2020, NASA submitted an official comment letter to the FCC regarding a request by AST & Science to operate a network of up 243 satellites at a 720 km orbit.[71] According to the report, the effect of a collision involving a satellite in any given constellation would be magnified, since the debris cloud could impede future operations at that orbital location. Therefore, the best way to manage a massive outpouring of satellites remains an open question.

Applying criticality of infrastructure systems to large satellite constellations is a complex issue, not only due to the complexity of their architecture, but because any adaptation of CI theory involves numerous factors that are interconnected in a global setting.[72] However, it is time for large satellite constellations, to be treated as a CSI. There is a need to develop applicable methods, tools and techniques to address risks and vulnerabilities associated with satellite constellations.

5 Critical Space Infrastructures Protection and Large Satellite Constellations

Whilst intentional attacks on satellites appear to be the most pressing, the list of unintentional threats dwarfs those that are malicious in intent. In this regard, issues related to the designation of mega-constellations as CSIs and protection against attacks, human-made space debris or space weather are perhaps amongst the most urgent in the space security agenda over the next five years.

In some ways, the international space law that was agreed to half a century ago is insufficient for addressing all of the issues associated with the deployment of a large number of satellites, including the protection of space infrastructures. Although regulators and organizations are already embroiled in discussions about how to cope with mega-constellations, it may be too late by the time any international regime emerges. The ESA is currently developing an automated system to help satellites avoid collision, but it may not be ready until 2023; by then, thousands of satellites will likely already have launched.[73]

The protection of satellite constellations as space infrastructures will require a paradigm shift with respect to the way that space missions are handled. In this context,

---

[71] Mike Wall, 'Planned Satellite Constellation Poses a Collision Threat' (*Space.com*, 7 November 2020) www.space.com/ast-science-satellite-constellation-collision-threat-nasa-warning.

[72] Alexandru Georgescu, 'Critical Space Infrastructures' in Kai-Uwe Schrogl et al. (eds), *Handbook of Space Security: Policies, Applications and Programs* (2nd edn, Springer Nature Switzerland AG 2020) 240.

[73] Automating Collision Avoidance (*ESA*, 22 October 2019) www.esa.int/Safety_Security/Space_Debris/Automating_collision_avoidance.

national homeland security policies must be revised as new generations of CSIs are built, including satellite constellations.

There are several options for increasing the resilience and security of satellite constellations, ensuring a space environment with manageable threats and enhancing the sustainability of CSIs.[74] They include the physical protection of satellites (e.g., resilient engineering), management of the orbital environment (e.g., post-mission disposal, STM and SSA) and multilateral initiatives (e.g., transparency and confidence-building measures and international agreements).

5.1 Resilience Through Physical Protection

In response to disruptive changes in the orbital environment due to the advent of satellite constellations, resilience was posited as one of the key approaches for maintaining superiority in space. As Peldszus mentioned, if it were evident to a potential adversary that a system would recover from attack or that any damage inflicted would either lead to a swift recovery or have limited repercussions on the overall capability afforded by a system or architecture, the calculus for an intentional attack would change.[75] Similarly, a resilient architecture would be able to withstand natural hazards in large satellite constellations. Beyond protective shielding (i.e., hardening), designing robust systems through resilient engineering is a key element of resilient CSIs.

The deployment of large constellations will both exacerbate the dynamics of the operational environment of orbits and offer new challenges and opportunities for the concept of resilience.

5.2 The Orbital Space Environment and Constellation Management

Reducing the quantity of space debris—and thus the danger of collisions—by removing inactive satellites and other defunct objects from orbits is one solution for the protection of satellite constellations. Debris remediation or 'active debris removal' can be realized in several ways, depending on the location of the debris.[76] In particular, objects in LEO can be forced to re-enter the Earth's atmosphere and

---

[74] Massimo Pellegrino and Gerald Stang, 'Space Systems and Critical Infrastructure' [2016] European Union Institute for Security Studies www.jstor.org/stable/resrep07091.6?seq=8#metadata_info_tab_contents.

[75] Regina Peldszus, 'Resilience of Space Systems: Principles and Practices' in Kai-Uwe Schrogl et al. (eds), *Handbook of Space Security: Policies, Applications and Programs* (2nd edn, Springer Nature Switzerland AG 2020) 140.

[76] Tanja Masson-Zwaan and Mahulena Hofmann, *Introduction to Space Law* (4th edn, Kluwer Law International BV, The Netherlands) 118.

burn up in a controlled re-entry. Unlike mitigation measures, which aim at reducing the number of objects, space debris remediation is designed to act against the consequences of orbital congestion with debris and aims at removing objects that are not functional anymore. However, such approaches present many technological and legal challenges with regard to mega-constellations, which must be examined.

Space traffic management concerns the planning, coordination and on-orbit synchronization of activities to enhance the safety, stability and sustainability of operations in the space environment. Constellation management is a subset of STM. In a 2019 study, different strategies were explored to enhance the level of automation so that more satellites will not translate into a proportional increase of managing effort. Proposed solutions included the optimization of automatic satellite tracking and automatic failure detection; thus, an operator would not need to manually check the satellite's status of health.[77]

Another element of STM is SSA, which enables the detection, tracking and identification of objects in outer space in order to protect space assets and predict and prevent collisions. Furthermore, an additional of STM is the establishment of 'rules of the road' for space traffic. The presence of such rules facilitates the attribution of liability in case of a collision in outer space, which is subject to faulty liability under the Liability Convention of 1972. Therefore, SSA is a key strategy for ensuring the continued resilience of large constellation infrastructures.

5.3 Multilateral Initiatives

Internationally, there have been a number of significant developments relevant to the mitigation of space debris. The most important of these is the Inter-Agency Space Debris Coordination Committee (IADC), an international forum of governmental bodies that include NASA, ESA and other space agencies, which adopted debris mitigation guidelines in 2002. Work by IADC also helped to inform the development of the Space Debris Mitigation Guidelines of the United Nations Committee on the Peaceful Uses of Outer Space (UNCOPUOS). Recent developments include the IADC issuance in 2017 of a 'Statement on Large Constellations of Satellites in Low Earth Orbit' as well as the adoption in 2019 by the UNCOPUOS 21 consensus guidelines for the 'Long-Term Sustainability of Space Activities'.[78]

At the same time as the multilateral discussions on UNCOPUOS started, the E.U. began a political initiative to develop a Code of Conduct for Outers Space Activities

---

[77] Giacomo Curzi et al., 'Large Constellations of Small Satellites: A Survey of Near Future Challenges and Missions' (2020) 7 (9) Aerospace www.mdpi.com/2226-4310/7/9/133.

[78] Annette Froehlich and Vincent Seffinga (eds), *The United Nations and Space Security: Conflicting Mandates Between UNCOPUOS and the CD* (Springer Nature Switzerland AG 2020) 114.

as a non-legally binding, international instrument aimed at building norms of responsible behaviour in space activities.[79] Furthermore, in 2010 a Group of Governmental Experts (GGE) on Outer Space Transparency and Confidence Building Measures (TCBMs) was established with the objective to improve international cooperation and reduce risks of misunderstanding and mistrust in outer space activities.[80]

Unlike GGE on space TCBMs and the COPUOS processes, the European Code of Conduct had no formal multilateral mandate and as a result this initiative stalled in 2015 due to a lack of diplomatic buy-in from other global players. However, although such instruments may be legally non-binding, they are politically binding and could contribute to the normative framework for outer space activities, as facilitate to predict and discipline state's behaviour with respect to maintaining the security of space.

6 Conclusion

Much of the world's CI is heavily dependent on space, specifically space-based assets, for its daily functioning. Essential systems—such as communications, navigation, climate monitoring and defence—all rely heavily on space infrastructures, including satellites, ground stations and data links.[81]

The increasing dependence on satellites places them firmly in the area of CI, whose disruption or destruction could have a devastating impact across a modern society's economy, security and sovereignty. This dependence poses a serious security dilemma for CIs services providers and governments.

However, the rapid increase in satellite constellations is a simmering crisis that is silently approaching the point of no return. The proliferation of satellites and orbital debris at altitudes less than 2000 km threatens the operations of existing and future satellites and the sustainability of high-value satellite orbits.

Like any other space infrastructure, satellites are vulnerable to natural or man-made threats. These vulnerabilities pose serious risks which requires a radical shift in the policies of international regulatory bodies towards the view of space assets as CSI.[82]

The protection of CIs has largely been approached under the guise of CIP, which concentrates upon the protection of the infrastructure and its components. While the security of the services provided by CI is obviously a concern, CIP advocates that

---

[79] Peter Martinez, 'Space Sustainability' in Kai-Uwe Schrogl et al. (eds), *Handbook of Space Security: Policies, Applications and Programs* (2nd edn, Springer Nature Switzerland AG 2020) 337.

[80] Ibid; George D. Kyriakopoulos, 'Security Issues with Respect to Celestial Bodies' in ibid 347.

[81] Meg King and Sophie Goguichvili, 'Cybersecurity Threats in Space: A Roadmap for Future Policy' (*Wilson Center*, 8 October 2020) www.wilsoncenter.org/blog-post/cybersecurity-threats-space-roadmap-future-policy.

[82] Alexandru Georgescu, 'Critical Space Infrastructures' in Kai-Uwe Schrogl et al. (eds), *Handbook of Space Security: Policies, Applications and Programs* (2nd edn, Springer Nature Switzerland AG 2020) 233.

the most effective method of maintaining those services is through the strengthening of the assets which carry them and the mitigation of any threats to said assets.[83]

With a threat-based approach to the security of space infrastructures, there is arguably an implicit intent to eliminate perceived threats instead of accepting that complete protection can never be guaranteed. For this reason, there has been a shift towards promoting resilience as an extension of the traditional protection efforts, suggesting that infrastructure security is beginning to be considered with regards to risk-based strategies instead of threat-based ones.[84]

Critical Infrastructure Resilience (CIR) diverges from CIP in numerous ways but particularly in that it concentrates upon the maintenance of services, rather than only the assets providing those services. Instead of rigid, and often expensive, physical protection, resilient strategies advocate a flexible and adaptive approach to services-security, whereby aspects of a systems are designed to be capable of redundancy in the event of a failure.[85]

In terms of security of CSI, the U.S. and E.U.'s approaches have been largely oriented towards protection. However, the focus on resilience within CIR discourses is intended to ensure that even though vulnerabilities within infrastructures may have been minimized through protection measures, damage or failures to parts of an infrastructure caused by unforeseen factors or risks should not affect the entire system.

John Tziouras is a Ph.D. candidate at Aristotle University of Thessaloniki, Faculty of Law. His research interest is in space security and space law. He obtained his LL.M. degree in International and European Legal Studies at Aristotle University of Thessaloniki. He is member of European Centre of Space Law. He is the founder of Spaceanalytica.eu.

---

[83] Phillip A. Slann, 'Anticipating Uncertainty: The Security of European Critical Outer Space Infrastructures' (2016) 35 Space Policy www.sciencedirect.com/science/article/abs/pii/S0265964615300163?via%3Dihub.

[84] Ibid.

[85] Regina Peldszus, 'Resilience of Space Systems: Principles and Practices' in Kai-Uwe Schrogl et al. (eds), *Handbook of Space Security: Policies, Applications and Programs* (2nd edn, Springer Nature Switzerland AG 2020) 140.

Satellite Constellations and the Sustainable Use of Outer Space

Long-Term Sustainability Guidelines as an Incentive Towards more Responsible Behaviour in Outer Space

Gina Petrovici

Abstract Over the past decades, the space community has experienced a paradigm change. With ongoing technological advancements, space applications have become indispensable for society. Among the results are a significant growth in the amount of space missions, and a similar increase in the number of public and private actors participating in space activities. Despite the impact of the COVID-19 pandemic in 2020, the space sector expanded and parts of constellations of thousands of satellites started to be launched. These constellations form a new trend in commercial space activities. The emergence of satellite constellations, combined with the increasing complexity of space operations, raises challenging questions regarding the risk of orbit congestion and interference. This contribution names some major risks that are severely jeopardising the long-term sustainability of space activities. While the classic *corpus iuris spatialis* is the starting point, a particular focus will be on the Long-Term Sustainability Guidelines, and their role as incentive for responsible behaviour of state and non-state actors and as a reliable starting point for developing rules of the road for outer space.

1 Introduction: Status Quo

Since the launching of the world's first artificial satellite "Sputnik 1" by the Soviet Union in 1957[1] and the US "Explorer" satellite shortly thereafter, technological advancements at a rapid pace led to a significant rise in the exploration and use of outer

---

[1] Heng (B.), Studies in international space law, Clarendon Press, 1997, 800; Vedeshin (L. S.), Dudykin (V. P.), «Preparation and launching in the USSR of the first artificial earth satellite», in Telecommunication Journal, Volume 44, 1977, pp. 477–481.

The views expressed are purely personal and do not necessarily reflect the view of any entity with which the author may be affiliated.

G. Petrovici (✉)
Leuphana Law School, Lüneburg, Germany

© The Author(s), under exclusive license to Springer Nature Switzerland AG 2021
A. Froehlich (ed.), *Legal Aspects Around Satellite Constellations*,
Studies in Space Policy 31, https://doi.org/10.1007/978-3-030-71385-0_6

space. Particularly in the last decade, the trend started to go in the direction of "NewSpace", the commercialisation and privatisation of space activities.[2] Previously, space activities had been predominantly conducted by States. A notable development in the recent years is the mass production of small satellites. Companies, such as SpaceX and Airbus/OneWeb are in the process of manufacturing thousands of broad-band communication satellites to be launched to an altitude of 1000–1200 km.[3] With these constellations,[4] space actors intend to enhance accessibility to telecommunication applications even in remote areas.

Nevertheless, this promising development has the potential of significantly increasing the likelihood of orbital collisions, leading to the rise of non-functional space objects in outer space further threatening functional space systems with the danger of triggering even more collisions and chain reactions (also known as the Kessler Syndrome[5]). The reason for this risk is that in addition to the approximately 3.000 currently operational satellites, authorities are monitoring around 20.000 pieces of orbital debris, these being limited to those pieces that are trackable[6]. A study on orbital debris collision in the context of constellations by Steel estimates a collision every twenty-five years of the OneWeb constellation and every twenty months of the SpaceX Starlink constellation.[7]

Despite the risks associated with the rapid growth in the amount of space objects in orbit due to satellite constellations, it is to the detriment of space sustainability that the smaller the satellites are in size, the less likely they are to be manoeuvrable. Also, if trackable at all, the fragments of collided small satellites are unlikely to be identified and monitored with the current techniques available, therefore placing the remaining parts of the constellation, other space missions and importantly the lives of astronauts onboard the International Space Station at risk. Even small pieces of space debris (less than the trackable >1 cm in size) can cause enormous damage considering the speed at which these fragments are moving. Such risk became reality on 02 September 2019, when the European Space Agency performed an evasive collision avoidance manoeuvre of its European Earth-observation satellite "Aeolus" approximately half an orbit before a potential collision with Space X's "Starlink 44"; one of the Companies' first sixty space objects (sent to space in a single launch) that are intended to form part of the satellite constellation for broad-band internet services.

---

[2] Hobe S and others, Cologne Commentary on Space Law (Carl Heymanns Verlag 2009) 14.

[3] For more information concerning the SpaceX Starlink constellation, see: www.starlink.com and for the Airbus/OneWeb Constellation, see: www.airbus.com/space/telecommunications-satellites/oneweb-satellites-connection-for-people-all-over-the-globe.html (all websites cited in this publication were last accessed and verified on 20 January 2021).

[4] For more information related to the topic of satellite constellations, see also: Annette Froehlich (ed.), Legal Aspects Around Satellite Constellations, in: Studies in Space Policy Vol. 19, 2019, Springer Nature Switzerland.

[5] Donald J. Kessler, "Collisional cascading: the limits of population growth in low Earth orbit" (1991) 11 Advanced Space Research, 63–66.

[6] Moriba Jah, Outer space is a mess that Moriba Jah wants to clean up, at: The Verge, www.theverge.com/science/22229792/space-orbital-collisions-risk-satellites-real-time.

[7] D. Steel, Assessment of the orbital debris collision hazard for the Low Earth Orbit satellites (2015).

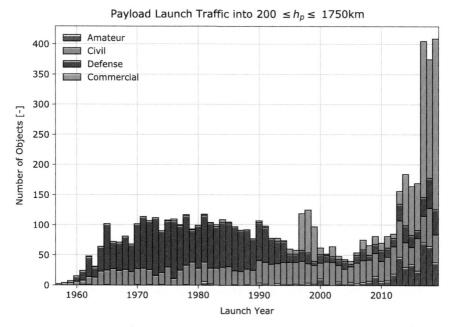

Fig. 1 Payload traffic in low earth orbit. *Source* ESA (2020), ESA's Annual Space Environment Report: Produced with the DISCOS Database, Retrieved from: https://www.sdo.esoc.esa.int/enviro nment_report/Space_Environment_Report_latest.pdf

As an outcome of this event, the European Space Agency underlined that collision avoidance manoeuvres of this kind may have to be conducted more frequently in the upcoming years, so that the manual monitoring of these operations "will likely become impossible".[8]

The following figure shows the payload launch traffic also in those altitudes of the low Earth orbit that are supposed to accommodate the aforementioned mega constellations (Figs. 1 and 2).

In support of the aforementioned argument that the increase in constellation launches is in line with the massive and rapid growth in space operations, a recent ESPI study on Space Traffic Management presents a forecast of satellite launches in the upcoming years.

It highlights the rapid increase in commercial space activities, characterised to a large extent by the launch of satellite constellations,[9] in this orbit and leads to the important note that even though the range of space actors is far more diverse than

---

[8] Mike Wall, "European Satellite Dodges Potential Collision with SpaceX Starlink Craft", on Space.com, www.space.com/spacex-starlink-esa-satellite-collision-avoidance.html.

[9] On the topic mega-constellations as a commercial space trend, see also: Marchisio (S.), «Opening panel: Space Law and Governance», at 10th United Nations Workshop on Space Law: Contribution of Space Law and Policy to Space Governance and Space Security in the 21th Century, Vienna, Austria, 2016.

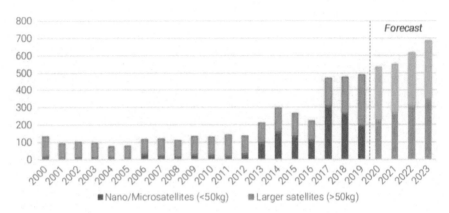

Fig. 2 Forecast of upcoming satellite launches. *Source* ESPI database, Euroconsult (2019), Satellites to be Built and Launched by 2027, via satnews. Retrieved from: www.satnews.com/strory.php?number=2091711277, SpaceWorks (2019). Nano/Microsatellites Market Forecast, 9th Edition

might have been foreseen more than sixty years ago, the interests of all these space actors are converging when it comes to the accessibility to space and the applications that can be linked to it. A central element to tackle these challenges is multilateral cooperation between state and non-state actors on many levels to reach a common understanding of best practices from a legal, regulatory, technical and economic perspective. This contribution sheds light on the first compendium of sustainability guidelines agreed to by consensus at the United Nations level (Sect. 1.4. below).

2 Corpus Iuris Spatialis

Following decades of lengthy negotiations, the international state community has seen the emergence of new codified legal instruments regulating the peaceful exploration and use of outer space for all. These negotiations within the United Nations fora led to several United Nations General Assembly Resolutions, numerous bilateral arrangements and guidelines introduced by International Organisations as well as implementation of these fundamental principles in domestic laws.[10] The *corpus iuris spatialis* consists of five main treaties, along with further non-binding decisions, drafted and concluded within the United Nations Committee on the Peaceful Uses of Outer Space (UN COPUOS), and serves as a highly relevant set of rules within public international law.[11]

---

[10]For an overview of domestic laws, see: www.unoosa.org/oosa/en/ourwork/spacelaw/nationalspacelaw.html.

[11]UNCOPUOS was established by the United Nations General Assembly in 1959, shortly after the launch of Sputnik 1: see United Nations General Assembly Resolution 1472 (XIV) on International co-operation in the peaceful uses of outer space (1959). It currently has 70 Members, which,

The 1967 Treaty on the Principles Governing the Activities of States in the Exploration and Use of Outer Space, including the Moon and other Celestial Bodies (Outer Space Treaty)[12] is the *magna carta* of space activities, setting out the general legal principles applicable to outer space activities, such as "the non-appropriation principle (also known as *res-communis omnium*)"; "the freedom of exploration and use"; a "liability regime applicable in the case of damage caused by space objects " and "the notification and registration of space activities with the United Nations".

These core legal principles are further elaborated in the subsequent multilateral treaties, id est the:

- 1968 Agreement on the Rescue of Astronauts, the Return of Astronauts and the Return of Objects Launched into Outer Space (Rescue and Return Agreement)[13];
- 1972 Convention on International Liability for Damage Caused by Space Objects (Liability Convention)[14];
- 1975 Convention on Registration of Objects Launched into Outer Space (Registration Convention)[15] and
- 1979 Agreement Governing the Activities of States on the Moon and other Celestial Bodies (Moon Agreement).[16]

In addition to the UN Space Treaties, the United Nations General Assembly adopted five sets of principles applicable to the exploration and use of outer space:

- 1963 Declaration of Legal Principles Governing the Activities of States in the Exploration and Use of Outer Space[17];
- 1982 Principles Governing the Use by States of Artificial Earth Satellites for International Direct Television Broadcasting[18];
- 1986 Principles Relating to Remote Sensing of the Earth from Outer Space[19];

---

according to UNCOPUOS, means that it is "one of the largest Committees in the United Nations." see: www.unoosa.org/oosa/en/members/index.html.

[12] Adopted by the General Assembly in its resolution 2222 (XXI), opened for signature on 27 January 1967, entered into force on 10 October 1967, 610 U.N.T.S. 205 (Outer Space Treaty).

[13] Adopted by the General Assembly in its resolution 2345 (XXII), opened for signature on 22 April 1968, entered into force on 3 December 1968, 672 U.N.T.S. 119 (Rescue Agreement).

[14] Adopted by the General Assembly in its resolution 2777 (XXVI), opened for signature on 29 March 1972, entered into force on 1 September 1972 961, U.N.T.S. 187 (Liability Convention).

[15] Adopted by the General Assembly in its resolution 3235 (XXIX), opened for signature on 14 January 1975, entered into force on 15 September 1976 1023, U.N.T.S. 15 (Registration Convention).

[16] Adopted by the General Assembly in its resolution 34/68, opened for signature on 18 December 1979, entered into force on 11 July 1984, 1363 U.N.T.S 3 (Moon Agreement).

[17] United Nations General Assembly Resolution 1962 (XVIII) on the Declaration of Legal Principles Governing the Activities of States in the Exploration and Uses of Outer Space.

[18] United Nations General Assembly Resolution 37/92 on the Principles Governing the Use by States of Artificial Earth Satellites for International Direct Television Broadcasting.

[19] United Nations General Assembly Resolution 41/65 on the Principles relating to Remote Sensing of the Earth from Outer Space.

- 1992 Principles Relevant to the Use of Nuclear Power Sources in Outer Space[20]; and
- 1996 Declaration on International Cooperation in the Exploration and Use of Outer Space for the Benefit and in the Interest of All States, Taking into Particular Account the Needs of Developing Countries.[21]

With the exception of those that might have reached customary law status, which has to be proven on a case-by-case basis,[22] these Resolutions are considered soft law; and have a non-legally binding character in the sense of Article 38(1) of the Statute of the International Court of Justice (ICJ).[23]

Even though the current intensity for governance of private space activities may appear to go beyond what the drafters of the traditional *corpus iuris spatialis* had foreseen, space activities carried out by private entities have been debated in the drafting process of the traditional space law instruments. Article VI of the Outer Space Treaty refers to activities of non-governmental entities in outer space. It provides a dual-system, whereby non-state actions in space are permissible under the prerequisite of state authority and control.[24] The particular impact on the sustainability of space activities by public and private actors is further developed in following sub-sections of this chapter.

3 Sustainability in the Corpus Iuris Spatialis and Beyond

Space activities have become indispensable for the well-being of humankind since they facilitate significant improvement of the standard of living through ensuring access to the benefits of space applications and information derived from space data in numerous areas, such as earth observation for the monitoring of climate change, navigation and telecommunication.[25] Conversely, these positive developments take

---

[20] United Nations General Assembly Resolution 47/68 on the Principles relevant to the Use of Nuclear Power Sources in Outer Space (Nuclear Power Source Principles).

[21] United Nations General Assembly Resolution 51/122 on the Declaration on International Cooperation in the Exploration and Use of Outer Space for the Benefit and in the Interest of All States, Taking into Particular Account the Needs of Developing Countries.

[22] See, for example, Ricky J. Lee and Steven Freeland, 'The crystallisation of general assembly space declarations into customary international law', (2004) 46 Proceedings of the Colloquium on the Law of Outer Space 122.

[23] Statute of the International Court of Justice (ICJ Statute), 1. U.N.T.S. 16. Following the opinion of international law scholars, Article 38(1) of the ICJ Statute lists the sources of international law, see: Antonio Cassese, International Law (2nd edn. 2005), 156.

[24] "States Parties to the Treaty shall bear international responsibility for national activities in outer space, including the Moon and other celestial bodies, whether such activities are carried on by governmental agencies or by non-governmental entities, and for assuring that national activities are carried out in conformity with the provisions set forth in the present Treaty."

[25] See also: Ulrike Maria Bohlmann, Gina Petrovici, Developing planetary sustainability: Legal challenges of Space 4.0. Global Sustainability 2 (2019), 1–11.

place in a complex net of legal, political and economic considerations. It is worth recalling that the exploration and use of outer space involves by its very nature elements that can be inherently threatening to the Earth and space environment. It is not a question, but a matter of fact that the international space community needs to find ways to ensure a safe and sustainable use of outer space for current and future generations since this is decisive for space actors and those benefitting from space activities.

3.1 Treaty Law Perspective

At the beginning, environmental considerations were often considered a hindrance to the emerging technological achievements that allowed States to conduct space activities.[26] However, even though environmental law was codified after the successful negotiation of the *corpus iuris spatialis,* the unanimously adopted 1961 General Assembly Resolution 1721 (xvi) on International Cooperation in the Peaceful Uses of Outer Space spotlighted the acknowledgement of the risks and benefits associated to an increase in launch activities. It also highlighted the potential that lays in international cooperation through expert fora, such as the United Nation bodies and the World Meteorological Organisations (WMO) with its related World Weather Watch (WWW) programme as well as the WMO-International Council for Science Global Atmospheric Research Programme.

Nevertheless, the *travaux préparatoires* of the Outer Space Treaty disclose that the drafters had taken the environmental impact of their activities in outer space into serious consideration, so that the Scientific and Technical Sub-Committee of the United Nations Committee on the Peaceful Uses of Outer Space recognized the necessity to ensure that potentially harmful interference must be avoided. The Committee on the Peaceful Uses of Outer Space based its resulting recommendations also on the specific proposals made by the Committee on Space Research (COSPAR[27]) Council on May 1964.[28] As a result of these deliberations in the drafting process, the space treaties and principles contain only implicit regulations and references.

Article IX, 2nd sentence represents the *sedes materiae* in the Outer Space Treaty relating to the protection of the space environment. It obliges States to conduct their exploration of outer space, including the moon and other celestial bodies; in such a manner as to "*avoid their harmful contamination and also adverse changes*

---

[26]Lesley I. Tennen, "Evolution of the planetary protection policy: conflict of science and jurisprudence?" (2004) 24 Advances in Space Research 2354.

[27]The Committee on Space Research (COSPAR) was founded in London in 1958 by the then International Council of Scientific Unions (ICSU), which is now the International Science Council (ISC). COSPAR is tasked to promote and enhance scientific research in space and provide a forum for discussion for the international scientific community. For more information see: https://cosparhq.cnes.fr/about/.

[28]Official Records of the General Assembly Eighteenth Session, Annexes, agenda item 28, documents A/5549, Annex II.

in the environment of the Earth resulting from the introduction of extra-terrestrial matter [...]" thereby, covering forward contamination of outer space and background contamination of the Earth environment. In addition, States are under an obligation to undertake appropriate international consultation if they have reason to believe that an activity or experiment planned by it or its non-governmental entities would cause potential harmful interference with activities of other States.

In addition, the Liability Convention establishes a system of absolute liability (Article II) to be applied in case of damage caused by a space object on the surface of the Earth, or to aircraft in flight, which could be interpreted as an Earth-centric stamp in the *corpus iuris spatialis*. Compared to that, the drafters of the Liability Convention developed a provision of fault-based liability (Article III) if the damage is caused elsewhere than on the surface of the Earth to a space object or to persons or property on board such a space object of another launching State. It is worth noting that Article I(a) contains a definition for the term "damage" as meaning "*loss of life, personal injury or other impairment of health; or loss of or damage to property of States or of persons, natural or juridical, or property of international intergovernmental organisations.*" Thus, not excluding environmental damage from those types of damage covered by the Liability Convention. However, it does not address any kind of environmental damage that is outside the surface of the Earth.

Moreover, other norms set forth in the United Nations space treaties make reference to the environmental considerations related to the exploitation of natural resources in outer space. Article 7 of the Moon Agreement, for example, refers to the so-called in situ resource utilization, which is a promising subject in the NewSpace era. It is widely discussed in relation to additive manufacturing (id est 3D printing), which has already been tested on the International Space Station and is intended to be used in the course of deep space exploration on the Moon and Mars. Since the Moon Agreement is the last of the five core space treaties, it was negotiated at a time when environmental considerations moved to the centre of international discussions. There is another novelty with regard to the principles of international environmental law and space law to be found in the Moon Agreement: the concept of intergenerational equity is at first space-specifically formulated in Article 7.1 of the Moon Agreement. Conversely, it is worth noting that the Moon Agreement so far has only gathered 18 State Parties—compared to 107 State Parties for the Outer Space Treaty, among which all major space-faring nations can be found.[29]

3.2 Broader Public International Law Perspective

Space law treaties do not only form part of public international law but its *magna charta*, the Outer Space Treaty specifically requests State parties to the treaty to comply with international law. Specifically, Article III of the Outer Space Treaty

---

[29]Status of international agreements relating to activities in outer space as at 1 January 2018, UN-A/AC.105/C.2/2018/CRP.3.

reaffirms that States Parties to the Treaty shall carry on activities in the exploration and use of outer space in accordance with international law, including international environmental law. Firstly, the 1972 Stockholm Declaration[30] set forth a significant legal statement with fundamental international principles for the protection of the environment.

In accordance with Principle 21 of the Stockholm Declaration, States are obliged in accordance with the Charter of the UN and the principles of international law, to take "*[...] the responsibility to ensure that activities within their jurisdiction or control do not cause damage to the environment of other States or of areas beyond the limits of national jurisdiction.*" As the second part of the sentence directly refers to areas beyond the limits of national jurisdiction and, considering that outer space is one of the areas beyond the limits of national jurisdiction, the outer space environment is imminently protected by this principle. The UN General Assembly Resolution 2996 (XXVII) 1972 asserts that Principle 21 and 22 of the Stockholm Declaration provide the fundamental rules governing the matter of environmental protection.

Furthermore, Principle 2 of the Declaration of the UN Conference on Environment and Development, adopted in Rio de Janeiro in 1992, and Article 3 of the 1992 Convention on Biological Diversity do repeat the previously presented notion. The ancient principle of *sic utere tuo ut alienum non laedas* (commonly known as "due diligence") as established by the International Court of Justice in the Corfu Channel Case[31], and further recalled in the Barcelona Traction Case,[32] provides for a liability regime based on a State's general duty of care. This notion has a great impact on the international state responsibility and thus also in the frame of environmental law. It is further advanced by Principle 15 of the Rio Declaration, which reads as follows:

> To protect the environment, the precautionary approach shall be widely applied by States according to their capabilities. Where there are threats of serious or irreversible damage, lack of full scientific certainty shall not be used as a reason for postponing cost-effective measures to prevent environmental degradation. Accordingly, if there is sufficient scientific evidence to establish the possibility of a risk of serious harm, States cannot justify their lack of action with the absence of a proof of harm.[33]

Additionally, this maxim was further affirmed by the International Court of Justice. In its Advisory Opinion on the Legality of the Threat or Use of Nuclear Weapons[34] the court declared: "*[...] the existence of the general obligation of States to ensure that activities within their jurisdiction and control respect the environment of other states or of areas beyond national control is now part of the corpus of international law relating to the environment.*"

---

[30] Declaration of the United Nations Conference on the Human Environment (16 June 1972) UN Doc A/CONF.48/14/Rev. 1 (1972 Stockholm Declaration).

[31] Corfu Channel, United Kingdom v Albania, Judgment, Merits, [1949] ICJ Rep 4.

[32] Barcelona Traction, Light and Power Company Limited (New Application, 1962), Belgium v Spain, Judgment, Merits, Second Phase, [1970] ICJ Rep. 3 (1970).

[33] Declaration of the United Nations Conference on the Human Environment (16 June 1972) UN Doc A/CONF.48/14/Rev.1 (1972 Stockholm Declaration).

[34] International Court of Justice Advisory Opinion on the Legality of the Threat or Use of Nuclear Weapons, [1996] ICJ Rep. 226, para. 29.

In addition, an emphasis was made on the responsibility and liability of states for transboundary harm in the Trail Smelter Arbitration.[35]

Throughout the past decades, efforts to formulate additional hard law instruments failed after lengthy negotiations. In contrast, the tendency is to establish soft law principles and guidelines to govern those aspects that are not yet fully covered by the traditional *corpus iuris spatialis*. Results are the COSPAR Planetary Protection Policy,[36] the United Nations Space Debris Mitigation Guidelines[37] and the Long-Term Sustainability Guidelines of Outer Space Activities.[38] The latter is discussed in the following section.

4 Long-Term Sustainability Guidelines of Outer Space Activities

The United Nations, as a multilateral diplomatic platform, has played a fundamental role in the exchange about and development of international environmental standards since the 1960s. The focus on sustainable aspects of life on Earth is among others reflected in the Human Development Report of 1990, which highlights the need for (i) national implementation and plans of action, (ii) adapting global goals to the national context, (iii) setting national priorities, (iv) ensuring cost effectiveness, and (v) adequate budget support.[39] The United Nations Millennium Declaration[40] is the result of a three-day Millennium Summit setting forth around sixty goals among which the environment also has a prominent role. These Millennium Development Goals (MDGs) form the basis for the United Nations Sustainable Development Goals adopted in 2015 by the United Nations General Assembly in its 2030 Agenda (A/RES/70/1).

With this in mind, the United Nations Committee on the Peaceful Uses of Outer Space adopted by the required absolute consensus of its then 92 member states, following an almost ten years process of international collaboration, the twenty-one guidelines for the long-term sustainability of outer space activities in 2019; these are referred to as the "Long-Term Sustainability Guidelines of Outer Space Activities or LTS Guidelines". This milestone achievement led to an accessible compendium of minimum standards addressing policy, regulatory, operational, safety, scientific,

---

[35] Trail Smelter Arbitration (US v. Canada) 1938/1941, R.I.A.A. 1905.

[36] The current version of the COSPAR Planetary Protection Policy, 20 October 2002, amended up December 2017, cosparhq.cnes.fr/assets/uploads/2019/12/PPPolicyDecember-2017.pdf.

[37] Adopted by the Scientific and Technical Subcommittee of UNCOPUOS at its 44th session in 2007, A/AC.105/890, para. 99, and endorsed by the United Nations General Assembly in Resolution 62/217 on international cooperation in the peaceful uses of outer space;

[38] Report of the 62nd session of the Committee on the Peaceful Uses of Outer Space (12–21 June 2019), UN General Assembly document A/74/20, Annex II. www.unoosa.org/res/oosadoc/data/doc uments/2019/a/a7420_0_html/V1906077.pdf.

[39] United Nations Development Report 1990, pp. 61–84.

[40] United Nations General Assembly Resolution 55/2 of 08 September 2000.

technical, international cooperation, capacity building and transparency aspects. The unique advantage of the guidelines and standards negotiated at the United Nations Committee on the Peaceful Uses of Outer Space the level is that these are concluded on a consensus basis. Thus, their acknowledgement by the international community is greater and they are more likely to be implemented on national level, allowing these legal norms to achieve the desired effects. It is now upon the States to identify ways to implement these guidelines on a national level and upon the international community to monitor these implementation practices to provide guidance, support and fora of exchange in this respect, as well as to raise awareness for the need for sustainable practices on Earth and in space. All of this with the aim to allow current and future generations to have an interference-free, safe and long-term sustainable access to outer space and the orbits surrounding Earth, which are a scarce resource after all.

4.1 Negotiation Process

Sustainability of outer space activities was raised for the first time in during the 50th session of the United Nations Committee on the Peaceful Uses of Outer Space (UN COPUOS) in 2005 when the Canadian delegate Mr. Karl Doetsch[41] presented a discussion paper on the role of the Committee in the upcoming 50 years.[42] Following that, during the 2006–2007 chairmanship of Mr. Gérard Brachet (France), the Chair highlighted the matter again by presenting a working paper identifying the long-term sustainability of outer space activities as one of the key challenges facing future peaceful uses of outer space. Moreover, it contained the first mention of a need to establish a Working Group in the form of the Scientific and Technical Sub-Committee to elaborate and assess its technical aspects. Further discussion throughout the 2008 and 2009 Sub-Committee sessions led to the establishment of a Working Group to address the long-term sustainability of outer space activities in 2010 under the chairmanship of Mr. Peter Martinez (South Africa). After agreement on the terms of reference, the scope and the methods of work in the 54th session of the Committee,[43] the Working Group was tasked to recommend a set of voluntary, thus non-binding, guidelines applied by States as commitments towards the long-term sustainability of

---

[41] Chair of the STSC from 2001 to 2003.

[42] See also: Karl Doetsch, Gérard Brachet, Peter Martinez, Kenneth D. Hodgkins, Richard Buenneke and Amer Charlesworth and Theresa Hitchens in Part I: The Multilateral Effort to Assure Space Sustainability in Space For The 21st Century—Discovery, Innovation, Sustainability (Michael Simpson, Ray Williamson and Langdon Morris eds., 2016.

[43] Committee on the Peaceful Uses of Outer Space, Report of the Committee on the Peaceful Uses of Outer Space Fifty-fourth session, pp. 51–57, Annex II. Terms of reference and methods of work of the Working Group on Long-Term Sustainability of Outer Space Activities of the Scientific and Technical Subcommittee, U.N. Doc. A/66/20 (2011) www.unoosa.org/pdf/gadocs/A_66_20E.pdf.

outer space activities throughout all phases of a mission life cycle, with the objective to boost safe space operations for generations.[44]

The traditional set of space treaties and principles governing the activities of states in the exploration and use of outer space provided the legal framework for the guidelines under development by the Working Group. While these recommend the implementation of the Long-Term Sustainability Guidelines of Outer Space Activities into national law (see 1.4.2. below), the Working Group excluded the task of developing new legally binding instruments.

The expected character of the Guidelines is described in the terms of reference as follows:

> (a) Create a framework for possible development and enhancement of national and international practices pertaining to enhancing the long-term sustainability of outer space activities, including, inter alia, the improvement of the safety of space operations and the protection of the space environment, giving consideration to acceptable and reasonable financial and other connotations and taking into account the needs and interests of developing countries.
>
> (b) Be consistent with existing international legal frameworks for outer space activities and should be voluntary and not be legally binding.
>
> (c) Be consistent with the relevant activities and recommendations of the Committee and its Subcommittees, as well as of other working groups thereof, United Nations intergovernmental organizations and bodies and the Inter-Agency Space Debris Coordination Committee and other relevant international organizations, taking into account their status and competence.[45]

Experts, including non-state entities, nominated by their national governments (State delegations) and intergovernmental bodies with permanent observer status in the Committee, served in four expert groups in an ad hominem capacity, not necessarily reflecting their national governments' position, to provide input as technical fora to the Working Group. These expert groups contributed with their own expertise and received further input, for example from the United Nations Group of Governmental Experts on Transparency and Confidence-Building Measures in Outer Space, the International Academy of Astronautics, the International Telecommunication Union, the United Nations Conference on Disarmament and the Inter-Agency Space Debris Coordination Committee, the European Space Agency and the World Meteorological Organisation as well as the European Organisation for the Exploitation of Meteorological Satellites, the Asia–Pacific Space Cooperation Organisation and the Group on Earth Observations.

Additional expert input was provided by the European Space Policy Institute and the Secure World Foundation. The four working groups had individual topics to cover, id est:

(A)   Sustainable space utilization supporting sustainable development on Earth;

---

[44]Peter Martinez, Development of an international compendium of guidelines for the long-term sustainability of outer space activities, in: Space Policy, Vol. 43, February 2018, pp. 13–17.

[45]Committee on the Peaceful Uses of Outer Space, Report of the Committee on the Peaceful Uses of Outer Space Fifty-fourth session, pp. 51–57, Annex II. Terms of reference and methods of work of the Working Group on Long-Term Sustainability of Outer Space Activities of the Scientific and Technical Subcommittee, U.N. Doc. A/66/20 (2011), www.unoosa.org/pdf/gadocs/A_66_20E.pdf.

(B) Space debris, space operations and tools to support collaborative space situational awareness;
(C) Space weather; and
(D) Regulatory regimes and guidance for actors.

Nevertheless, joint meetings served as a platform for exchange about those topics that are interconnected and thus cannot be strictly categorised.[46] These expert groups then provided their recommendations in the form of thirty-one candidate guidelines and additional matters for further consideration to the Working Group in 2014.[47]

In the aftermath of this initial process, the Working Group started with the drafting process of the guidelines that ultimately resulted in a total of twenty-one. At this stage, the negotiation was conducted at the level of the Committee, and States took the occasion to propose draft guidelines for the Working Group's consideration. In a lengthy negotiation process, which was also due to the rapidly increasing number of memberships to the Committee and the growing interest by numerous States in the subject matter, agreement on the first twelve guidelines was achieved.[48] Following this, agreement was reached on nine additional guidelines and the preamble in the course of the 55th session of the Scientific and Technical Sub-Committee in February 2018. This agreement was of utmost importance, not only for the general progress in the negotiating the first internationally accepted compendium of best practices for space sustainability, but also as it set forth the definition of the term space sustainability as:

> the ability to maintain the conduct of space activities indefinitely into the future in a manner that realizes the objectives of equitable access to the benefits of the exploration and use of outer space for peaceful purposes, in order to meet the needs of the present generations while preserving the outer space environment for future generations.[49]

Thus, reflecting the notion of sustainability in the United Nations Brundtland report.[50]

---

[46] Peter Martinez, Space Sustainability, in: International Space Security Setting, Handbook of Space Security, Kai-Uwe Schrogl (ed.), 2020, pp. 257–272.

[47] See: Report of LTS expert group A—Sustainable space utilization supporting sustainable development on Earth, COPUOS session document A/AC.105/C.1/2014/CRP.13; Report of LTS expert group B—Space debris, space operations and tools to support collaborative space situational awareness, COPUOS session document A/AC.105/C.1/2014/CRP.14; Report of LTS expert group C—Space weather, COPUOS session document A/AC.105/C.1/2014/CRP.15; Report of LTS expert group D—Regulatory regimes and guidance for new actors in the space arena, COPUOS session document A/AC.105/C.1/2014/CRP.16.

[48] United Nations Committee on the Peaceful Uses of Outer Space (2016) Report of the fifty-ninth session, The first batch of agreed guidelines is contained in the Annex to the report of the 59th session of COPUOS in 2016, contained in UN document A/71/20.

[49] United Nations Committee on the Peaceful Uses of Outer Space (2018) Working Group on the Long-Term Sustainability of Outer Space Activities: preambular text and nine guidelines, conference room paper by the Chair of the Working Group on the Long-Term Sustainability of Outer Space Activities. UN document A/AC.105/C.1/2018/CRP.18/Rev.1.

[50] United Nations General Assembly (1987) Report of the World Commission on Environment and Development, Annex Our Common Future, UN document A/42/427.

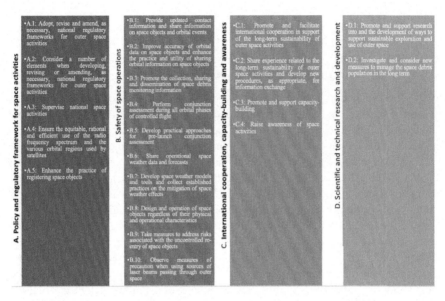

Fig. 3 The twenty-one long-term sustainability guidelines of outer space activities

However, the negotiation process did not conclude in 2018 since, despite the successful agreement of twenty-one guidelines,[51] the Working Group could not reach the required consensus on an additional seven guidelines before the mandate of the Working Group expired in June 2018. The main stumbling blocks have been the diverse opinions as related to the way forward. In addition, some States did not agree to detach the remaining seven guidelines from the previously agreed twenty-one guidelines.

Despite the end of the Working Groups' mandate, the discussion on the matter continued in 2019 with the result of the adoption of the preamble and the following twenty-one guidelines at the 62nd session of the Committee (Fig. 3)[52]:

In this context, it was strongly recommended that States and international intergovernmental organisations take steps to implement the guidelines to the greatest extent feasible and practicable. It is worth noting that such implementation would be on a voluntary basis.

---

[51] United Nations Committee on the Peaceful Uses of Outer Space (2018) Draft guidelines for the long-term sustainability of outer space activities, working paper by the Chair of the Working Group on the Long-Term Sustainability of Outer Space Activities, UN document A/AC.105/C.1/L.367.

[52] United Nations General Assembly (2019) Report of the Committee on the Peaceful Uses of Outer Space, sixty-second session (12–21 June 2019), UN document A/74/20, Annex II.

4.2 Implementation Process

It is clear that the Long-Term Sustainability Guidelines of Outer Space Activities will unfold their full potential and effect if implemented to the widest extend feasible by States and international intergovernmental organisations, and if the international community raises awareness for the need to ensure a sustainable use of outer space during the entire mission life-cycle. Once the guidelines are implemented on a national level, they will also provide the framework for non-state space actors and influence their mission planning and the sustainable conduct of space activities in general. In addition, these guidelines may serve as starting point towards sustainable interoperable common exploration infrastructures and the concretisation of rules of the road in space. The Long-Term Sustainability Guidelines of Outer Space Activities could be interpreted as a tool to incentivise responsible behaviour in space.

State and non-state actors already present the first implementation efforts. Be it in the form of national laws, technical instruments, such as ASTRIAGraph[53] and multilateral initiatives, such as the World Economic Forum's Space Sustainability Ranking,[54] with the aim to foster transparency and the free flow of information of orbital data to all space actors.

Importantly, international cooperation was and still is key to achieve progress for the benefit of space actors and beneficiaries from space activities. As a result, the Long-Term Sustainability Guidelines of Outer Space Activities not only build upon the notion of international cooperation but also contain numerous references to it themselves. Comparable with the Brundtland Commission's report, the *corpus iuris spatialis*, thus also the Long-Term Sustainability Guidelines of Outer Space Activities that are based on the traditional space law treaties, request States to take into particular consideration the needs of developing countries. Therefore, state and non-state actors are also asked to support developing countries in the implementation of the guidelines.

4.3 LTS 2.0

The Scientific and Technical Subcommittee decided to establish a new Working Group under the long-term sustainability agenda item (A/74/20, para 165), which still needs to select its bureau and a precise work plan to fulfil its tasks related to the:

(i) identification and study of challenges as well as potential development of new guidelines in this respect (including all those not yet agreed to in the previous negotiation process);

---

[53] Examples are multifaceted, one of them is ASTRIAGraph, for more information see: https://astria.tacc.utexas.edu/AstriaGraph/.

[54] For more information see: www.weforum.org/projects/space-sustainability-rating.

(ii) sharing experiences from the implementation of the Long-Term Sustainability Guidelines of Outer Space Activities; and
(iii) raising awareness and capacity building in particular among merging space nations and developing countries.[55]

Bringing the LTS topic forward may, due to rapidly advancing technological capabilities and the diverse interest of the multiple state actors in the Committee, give rise to challenges as concerns the required consensus-based decision-making processes.

The upcoming years will prove how States tend to implement the guidelines and which new requirements emerge in the exchange of state and non-state actors with a focus on strategic, legal, economic and technical aspects of space missions. Satellite-constellations, as one of the most prominent trends at the moment, as well as topics like on-orbit servicing[56] and space traffic management are eminently linked to the sustainable use and exploration of outer space and will move to the centre of discussions in the upcoming years. Accordingly, the likelihood of influencing the set of Long-Term Sustainability Guidelines of Outer Space Activities is very high.

The Long-Term Sustainability Guidelines of Outer Space Activities will remain a "living document" to adapt in the form of a "continued institutionalised dialogue"[57] to changes required to keep outer space a peaceful, sustainable and promising area of human exploration and use.

5 Space Traffic Management

As already emphasised on previous occasions in this Chapter, there is a need to ensure that, even though the orbits are becoming congested and participation in space activities is becoming increasingly diverse, access to and safe operation in outer space is granted. While on Earth rules of the road exist to regulate the traffic independent from the number of vehicles on the street, and an air-traffic management system ensures the safety of air flight every day, the *corpus iuris spatialis* is so far lacking a traffic regime for outer space even though it is clear that the traffic in outer space is rapidly exceeding the current launch rates. The term "space traffic" itself was already discussed almost forty years ago, then again ten years later, and elaborated on in depth in the International Academy of Astronauts study of 2006.[58] On the international level, the development of a regulatory approach to manage space traffic moved to

---

[55] Ibid.

[56] For more information related to the topic of On-Orbit servicing, see: Annette Froehlich (ed.), On-Orbit Servicing: Next Generation of Space Activities, in: Studies in Space Policy Vol. 26, 2020, Springer Nature Switzerland AG.

[57] See the Preamble of the Long-Term Sustainability Guidelines of Outer Space Activities, UN document A/74/20, Annex II.

[58] Corinne Contant-Jorgenson, Petr Lála, Kai-Uwe Schrogl, Cosmic Study on Space Traffic Management, Paris, International Academy of Astronautics (IAA) 2006.

the forefront of discussions by and between States, international intergovernmental organisations and non-state actors. According to the findings presented in the Space Traffic Management Roadmap, the notion of Space Traffic Management (STM) can be defined as: *"the set of technical and regulatory provision for promoting safe access into outer space, operations in outer space and return from outer space to Earth free from physical and radio-frequency interference"*.[59] Consequently, complying with the universal freedom principle of Article I Outer Space Treaty, States are not permitted to prevent or complicate the free access of other States to outer space.[60]

Despite the need to establish a Space Traffic Management regime,[61] so far, no consensus has been reached on concrete formulations at international level. It is worth noting that such a regime would require an assessment of the entire mission-life-cycle, so as to include for example pre-launch notifications, and it would also require a large degree of harmonisation of existing space laws on national and international levels.[62] In summary, a solid space traffic management system would rely on an interdependent set of rules based on legal and strategic, but importantly, also technical requirements and information. The latter being formed by data sets and algorithms to ensure the precision and efficacy of space missions. Data is already providing information about space objects at a given moment in time in a certain orbit and can mostly indicate the entity in charge of controlling a respective object.

An umbrella for this kind of measurements is Space Situational Awareness (SSA) data. These data sets are collected and analysed by various entities, some examples are listed here below.

Since June 2018, the United States Space Policy Directive-3 (83 FR 28969) sets a formal setting for the SSA and Space Traffic Management strategy in the United States. In light of this, the 18-Space Control Squadron (18-SPCS), formerly known as the Joint Space Operation Center (JSpOC), for the United States Strategic Command's (USSTRATCOM) space situational awareness programme, continues to collect and share data for the low Earth and geostationary orbits with foreign governments and private entities. In this framework, the German Space Operations Center (GSOC) of the German Aerospace Center is one of the data recipients. The latter is itself currently working on a dedicated sensor network "Small Aperture Robotic Telescope Network (SMARTnet)" for the monitoring of the geostationary orbit.

Moreover, the European Union established the Space Surveillance and Tracking Support Framework, also known as EU-SST, with Decision 541/2014/EU of the European Parliament and the European Council. The European Union Member States in the EU-SST Consortium are represented by their national designated entities, id est: Germany (DLR), France (CNES), Italy (ASI), Poland (POLSA), Portugal (PT

---

[59] Kai-Uwe Schrogl, Corinne Contant-Jorgenson, Jana Robinson, Alexander Soucek, Space Traffic Management—Towards a Roadmap for Implementation, Paris, IAA, 2018.

[60] Manfred Lachs, Tanja Masson-Zwaan, and Stephan Hobe, The Law of Outer Space an Experience in Contemporary Law-Making (Martinus Nijhoff Publishers 2010) 45.

[61] For more information related to the topic of space traffic management, see also: ESPI Report 71, Towards a European Approach to Space Traffic Management—Full Report, January 2020.

[62] Corinne Contant-Jorgenson, Petr Lála, Kai-Uwe Schrogl, Cosmic Study on Space Traffic Management, Paris, International Academy of Astronautics (IAA) 2006.

MoD), Romania (ROSA), Spain (CDTI) and the United Kingdom (UKSA). Together with the European Union Satellite Centre (SatCen), the Consortium has ever since collaborated to foster European Space Surveillance and Tracking capabilities.

Another example, is the European Space Agency's Space Safety programme (S2P), which build upon the agency's former optional Space Situational Awareness programme and includes activities related to space debris, space weather and planetary defence. Its Member States agreed at the last Ministerial Council in 2019, to place a service contract for the first space mission to remove a space debris from orbit. The resulting "ClearSpace-1" mission is scheduled for launch in 2025 and will benefit from the findings obtained in the Active Debris Removal In-Orbit Servicing (ADRIOS) project. As the Agency's Director General underlined:

> Imagine how dangerous sailing the high seas would be if all the ships ever lost in history were still drifting on top of the water. That is the current situation in orbit, and it cannot be allowed to continue. ESA's Member States have given their strong support to this new mission, which also points the way forward to essential new commercial services in the future.[63]

An approach needs to be identified in order to handle these massive data sets and their interpretation for an effective data and thus space traffic management.

Nevertheless, as concerns the legal and policy aspects, soft law instruments, like the Inter-Agency Space Debris Mitigation Guidelines and the Long-Term Sustainability Guidelines of Outer Space Activities, encompass principles of a Space Traffic Management. In the long-term, the freedom of exploration and use of outer space (Article I Outer Space Treaty) can only be complied with by conducting sustainable space missions. Congestion of orbits could lead to a violation of that principle since it endangers humankind's continuous ability to launch and operate in orbits (as a scarce resource). Therefore, multiple state and non-state actors are already elaborating standards and best practices for the launch, operation and end-of-life cycle of space objects to limit the number of satellites remaining in outer space after operability. Considering that the multilaterally agreed twenty-one Long-Term Sustainability Guidelines of Outer Space Activities are viewed as a living document that may be further expanded in the course of an institutionalised dialogue to react to technological advancements and other urging developments in the space sector, it is likely that a stronger connection could be made between the guidelines and rules of the road. Ultimately, both concepts aim to ensure an ongoing and un-interrupted access to space for all.

---

[63] ESA, ESA commissions world´s first space debris removal (2019), www.esa.int/Safety_Security/Clean_Space/ESA_commissions_world_s_first_space_debris_removal.

6 Outlook

As the relevance of space activities increasingly grows, the law governing these activities is of utmost importance for space actors and indirectly also for space beneficiaries. While the benefits space applications provide to society are undoubted, this development requires a multilateral approach to govern multifaceted space activities and thus, space traffic in general. It is clear that cooperation remains a core element of the peaceful and sustainable uses of outer space.

The previously existing *corpus iuris spatialis* touches rather indirectly upon environmental protection, but does not provide a comprehensive legal framework for the protection from forward and backward contamination. General obligations relating to the environmental aspects of the exploration and use of outer space that are found in the five United Nations space treaties or related principles need to adapt to recent scientific, technological, strategic and industrial developments, such as the launch and operation of satellite constellations that are providing chances, but also causing risks that need to be thoroughly taken into account on international level. State Parties to the Outer Space Treaty are bound by its obligations, and have to continuously supervise and control their activities, irrespective of whether they are conducted by the government or private actors (Article VI Outer Space Treaty). It is in the interest of those States to provide concrete norms of behaviour for the life-span of missions in order to comply with their obligation under public international law. In the same vein, it should be in the interest of non-space actors to contribute to the sustainable use of outer space in order to allow the expansion of space activities continue in a responsible manner, so that current and future generations can benefit from the advantages that space applications provide to society, being it in the field of telecommunications, earth observation for the monitoring of climate change and global pandemics or in many other areas that space activities already serve today.

As concerns the international fora, environmental considerations have increasingly gained relevance for the space community in the past ten years. The agreement on a set of twenty-one multidisciplinary Long-Term Sustainability Guidelines of Outer Space Activities by the United Nations Committee on the Peaceful Uses of Outer Space spotlights the increasing awareness of state and non-state actors in this respect. Moreover, these guidelines can be seen as an incentive for further actions in this respect and the continuous improvement and required adaption of standards to the fast developments in the space sector. Even though the guidelines are a compendium of non-binding soft law principles, they are also the result of thoughtful deliberations and multilateral consensus-based decision making. Thus, the guidelines will serve as a valuable compendium to face the challenges imposed by new trends, such as satellite constellations, and they are very likely to influence other regulatory frameworks (binding and particularly non-binding), such as rules of the road for outer space activities. It is clear that mutual trust, enhanced capabilities, clear legal frameworks and strategic guidelines, as well as multilateral cooperation in these areas, are key for a successful way forward.

Gina Petrovici, LL.M, joined the DLR Department of UN-Affairs after completing a DLR national young graduate trainee programme in the ESA Legal Services Department. Prior to that, she worked as a research assistant, (junior) lecturer and MLMC moot court coach in the field of international law and space law. She completed a six months traineeship in ESA's Strategy Department in summer 2018 after previously working already in DLR's Department of UN Affairs in 2018/2019. Gina holds a Master of Laws from the University of London (UCL and Queen Mary University) in External Relations of the European Union Law, International Trade Law, Telecommunications Law as well as Intellectual Property Rights. She has published contributions in books, journals and studies with a focus on various highly topical issues such as NewSpace, earth observation, STM and data policy. She is an active member of IISL, ECSL, WIA-E and SGAC.

Environmental Principles of Corporate Social Responsibility and Their Application to Satellite Mega-Constellations in Low Earth Orbit

S. Hadi Mahmoudi and Aishin Barabi

Abstract Mega low orbit satellite constellations are one of the largest commercial projects that will directly affect the daily lives of people. The most important achievement of these projects is *inter alia* the provision of internet access that could potentially cover the entire planet. However, despite these positive achievements, the challenges and environmental concerns pertaining to these projects should not be overlooked. These projects are typically carried on by private commercial companies. Furthermore, extensive international efforts have been made to ensure that these companies, in addition to their legally binding obligations, will behave in a socially responsible manner, which is referred to as "Corporate Social Responsibility". Although the environmental dimension of corporate social responsibility is primarily considered to be the effects of the business on this planet, the interpretation used in this essay incorporates environmental effects in any place in which companies operate. While examining the principles governing corporate social responsibility, this article argues that the international space law is a good ground for the implementation of these principles by commercial companies operating in the mega satellite constellations industry.

1 Introduction

The structure of the legal regime of outer space is predominantly developed for governmental and inter-governmental relations. Since the 1990s, however, with the emergence and growth of the private sector in the space industry, outer space, and its legal framework have faced a transition from governmental and public space activities

---

S. H. Mahmoudi
Shahid Beheshti University, Tehran, Iran
e-mail: h_mahmoudi@sbu.ac.ir

A. Barabi (✉)
Space Law and Policy, Beihang University, Beijing, China
e-mail: aishin.barabi@gmail.com

© The Author(s), under exclusive license to Springer Nature Switzerland AG 2021
A. Froehlich (ed.), *Legal Aspects Around Satellite Constellations*,
Studies in Space Policy 31, https://doi.org/10.1007/978-3-030-71385-0_7

to privatized and commercialized ones.[1] Space companies, mostly in space-faring countries, have planned and operated various commercial space projects seeking multiple goals and achievements including satellite telecommunication, data transmission, remote sensing of the Earth, commercial space transportations, tourism, the exploitation of extraterrestrial resources, and so on.

The recent development in the space industry that has brought concerns to the international community, is mega-constellations of satellites. Satellite constellations are defined as a group of satellites that operate at a specific altitude to provide wider coverage and follow the same task that can vary from global positioning, navigation, and disaster monitoring to communications. According to NASA, Satellite constellations are defined as "a group of satellites with a common purpose. The spacecrafts are located in particular configurations to accomplish the full mission."[2] It is also defined as a group of similar satellites, of a similar type and function, designed to be in similar, complementary, orbits for a shared purpose, under shared control.[3] The large-scale satellite constellations, also known as mega-constellations of satellites, include thousands of satellites planned to be deployed in different orbits namely Low Earth Orbits (from 160 to 2000 km, hereinafter LEO), Medium Earth Orbit (10000 ~ 20000 km, MEO), and Geostationary Orbits (altitude 35790 km, GEO). The most popular region for satellite mega-constellations is the LEO region due to the cheaper and less complicated technologies required for launching satellites.

In most cases, the operators of such satellite constellations are private companies rather than governmental space agencies. That is to say, the governmental nature of the existing legal framework governing the space activities can hardly deal with the problems arising from the new conundrum and challenges of satellite mega-constellations. Indeed, other legal mechanisms should be taken into consideration when it comes to these constellations bearing in mind that the volume of mega constellation projects is increasing rapidly. That is why this article seeks to find another framework that can be utilized for the issue of satellite constellations. Therefore, considering the mega-constellations projects as a business that is run by large, private space companies, the relevance of the principles of corporate social responsibility (hereinafter CSR) and responsible business has been taken into account. Although the mega-constellation of satellites will bring many benefits to the people worldwide, the drawbacks and disadvantages of these businesses should not be overlooked.

Today, one of the most efficient solutions predicted for running a safe and sustainable business is the core principles of CSR that aim to eliminate the negative effects

---

[1] P. Frankowski, Outer space and private companies consequences for global security, Politeja-Pismo Wydziału Studiów Międzynarodowych i Politycznych Uniwersytetu Jagiellońskiego, Vol 14, No. 50 (2017): 131–147.

[2] J. Howard, O. Dipak, S.S. Danford, Best Practices for Operations of Satellite Constellations, https://ntrs.nasa.gov/api/citations/20080039173/downloads/20080039173.pdf [all websites cited in this publication were last accessed and verified on 20 January 2021] (2006): 2.

[3] L. Wood, Satellite constellation networks, In *Internetworking and Computing over Satellite Networks*, Springer, Boston, MA, chapter 2 (2003): 13–34.

of businesses on the environment and society.[4] As mentioned before, the mega-constellations of satellite projects have a vast range of applications and activities that need to be regulated either by competent authorities through national laws or by private companies themselves through self-regulation. For this reason, the companies who execute constellations should be aware of the principles of CSR and responsible business and apply them to their business by the means of self-regulation.

Based on the previous researches that have been done in the field of mega-constellations and their negative effects such as the high risk of space debris generation and the excessive occupation of the orbital slots, it is assumed that the operating companies do not adhere to CSR principles. The main goal of this article is not only to encourage private space companies but also States to incorporate the principles of CSR into their projects and national space legislation to bring sustainability to outer space and their businesses.

For this purpose, first, the core principles governing CSR will be examined then the receptiveness of the current international legal regime of outer space to CSR principles will be discussed. The last and most important part of this research applies the CSR principle to satellite mega-constellations projects currently taking place, such as Starlink and OneWeb as examples to examine whether these business projects adhere to the aforementioned principles. Hopefully, this research will persuade the space companies to comply with the principles of CSR and responsible business.

2 A Brief Explanation of Corporate Social Responsibility

It is necessary to briefly discuss the history and definition of corporate social responsibility. Various definitions of corporate social responsibility have been identified. Naturally, this essay chooses a definition that can be effectively applied with respect to outer space activities. Moreover, many international efforts have been made by various international bodies to formulate these principles. We will review these documents as far as environmental aspects are concerned.

2.1 History and Definition

The impacts of business activities of the companies on the world around them was a topic of discussion since the industrial revolution. However, the emergence of the notion of CSR in the academic literature goes back to the 1930s when the adverse

---

[4]D. Katamba, Z. Christoph, H. David, T. Charles, Principles of Corporate Social Responsibility (2012).

impacts of the activities of businesses led to social movements and academic discussions.[5] Corporate responsibility towards society and the environment was specifically defined in the 1950s and 1960s by growing academic research and it can be regarded as the beginning of the modern notion of CSR.[6] In 1953, Bowen defined CSR as "the obligation of businessmen to pursue those policies, to make those decisions, or to follow those lines of action which are desirable in terms of the objectives and values of our society".[7] This series of events and academic researches around CSR resulted in growing concern and awareness not only among citizens but also among scholars who assisted the development of the modern notion of CSR in its modern way.

The 1990s is known as the era of internationalization of the concept of CSR, due to some chain of events, such as the establishment of the European Environment Agency in 1990, the United Nations (hereinafter UN) summit on the Environment and Development leading to the Rio Declaration on Environment and Development and, the adoption of the Kyoto Protocol in 1997.[8] Following the creation of international bodies, the reputational competition among multinational corporations for environmental and social considerations increased. The businessmen and corporations on a global scale started to recognize the significance of being socially and environmentally responsible as a way of achieving a better business reputation and consequently ensuring the sustainability of their businesses.

In 2000, the international bodies embarked on the journey of defining, regulating, and implementing CSR seriously. In July 2000, the UN Global Compact was established following the speech of the UN secretary-general, Kofi Annan, at the World Economic Forum in 1999 about the importance of CSR.[9] The primary goal of UNGC was to define a lacking instrument that helps the integration of human rights, social and environmental values into business activities.[10] The stream of CSR actions was not limited to the UN, other regions in the world also began to publish frameworks in this regard. In 2001, the European Commission published a green paper named Promoting a European Framework for Corporate Social Responsibility, and the European Strategy on CSR in 2002. In its 2001 document, the European Commission defined CSR as "A concept whereby companies integrate social and environmental concerns in their business operations and in their interaction with their stakeholders on a voluntary basis. Corporate social responsibility provides the foundations of an integrated approach that combines economic, environmental and social interests to

---

[5] A. B. Carroll, A history of corporate social responsibility: Concepts and practices, *The Oxford handbook of corporate social responsibility* (2008): 19–46.

[6] A. Latapí, J. Lára D. Brynhildur, A literature review of the history and evolution of corporate social responsibility, *International Journal of Corporate Social Responsibility* Vol 4, No. 1 (2019): 1.

[7] H. Bowen, *Social responsibilities of the businessman*, University of Iowa Press (2013).

[8] Ibid. *supra note 7*.

[9] United Nations Global Compact, UN History—A giant opens up. http://globalcompact15.org/report/part-i/un-history-a-giant-opens-up.

[10] Business as a force for good, www.unglobalcompact.org/what-is-gc/mission.

their mutual benefit."[11] However, in its renewed version in 2011 it clarified that CSR is "the responsibility of enterprises for their impacts on society".[12]

Moreover, with the occurrence of some relevant events which highlighted the importance of CSR including the Deepwater Horizon oil spill in the Gulf of Mexico in 2010[13] and the collapse of the Dhaka garment factory in 2013,[14] politicians, social activists, stakeholders, and consumers began to deeply understand how important it is for businesses to act in compliance with CSR.

2.2 Core Environmental Principles of CSR

When it comes to principles of CSR, three main internationally recognized instruments namely, the UN Guiding Principles on Business and Human Rights (hereinafter UNGPs),[15] the Guidelines of the Organization for Economic Co-operation and Development for Multinational Enterprises (hereinafter OECD guidelines), and the UN Global Compact principles (hereinafter UNGC principles) come to mind. Although the UNGPs are known as the building blocks of a responsible business, the environmental principles of CSR are mainly mentioned in UNGC principles and OECD guidelines which are in line with the UNGPs. Thus, the main focus of this section is on these instruments.

2.2.1 The UNGC Principles

The UNGC, founded in 2000, is the largest initiative in the world that provides a non-legally binding framework for sustainable business conduct. Three years after its establishment, the UNGC principles were launched containing 9 principles regarding human rights, labor, and the environment. Later in 2004, another principle about corruption was added to the 9 previous principles. Principles 7, 8 and 9 of this initiative address the issue of the environmental considerations in operating responsible business conducts. The principles are as follows:

---

[11] Promoting a European framework for Corporate Social Responsibility (2001).

[12] European Commission, A renewed EU strategy 2011–14 for corporate social responsibility, Communication from the Commission to the European Parliament, the Council, the European Economic and Social Committee and the Committee of the Regions (2011).

[13] Pallardy R, Deepwater Horizon oil spill, Encyclopedia Britannica, www.britannica.com/event/Deepwater-Horizon-oil-spill (2 November 2020).

[14] Bangladesh factory collapse toll passes 1000, BBC News, www.bbc.com/news/world-asia-22476774.

[15] J. Ruggie, Palais Des Nations, Guiding principles on business and human rights: Implementing the UN "Protect, Respect and Remedy" Framework, *Report of the Special Representative of the Secretary General on the issue of human rights and transnational corporations and other business enterprises* (2011). United Nations, 'Guiding Principles on Business and Human Rights Guiding Principles on Business and Human Rights'.

Principle 7: Businesses should support a precautionary approach to environmental challenges;

Principle 8: Businesses should undertake initiatives to promote greater environmental responsibility;

Principle 9: Businesses should encourage the development and diffusion of environmentally friendly technologies.[16]

These CSR principles will be further used and applied for different issues regarding mega satellite constellations.

2.2.2 The OECD Guidelines

The OECD guidelines were first published in 1976, along with and as an annex to the OECD Declaration on International Investment and Multinational Enterprises.[17] In 2011, the guidelines were updated adding many important changes. As mentioned in its preface, the guidelines provide recommendations and standards from adhering governments to their multinational enterprises, as a way of helping businesses to become socially and environmentally responsible.

The guidelines include different sections namely general policies, disclosure, human rights, employment and industrial relations, environment, combating bribery, bribe solicitation and extortion, consumer interests, science and technology, competition, taxation, and implementation procedures. Section 6 of the guidelines is concerning environmental issues. This section mentions that:

> Enterprises should, within the framework of laws, regulations and administrative practices in the countries in which they operate, and in consideration of relevant international agreements, principles, objectives, and standards, take due account of the need to protect the environment, public health and safety, and generally to conduct their activities in a manner contributing to the wider goal of sustainable development.[18]

Since the operation phase of the mega-constellations is in the outer space environment, and for the purpose of applying environmental principles to the orbital environment in order to help its preservation, the phrase "in countries in which they operate" is broadly, and in good faith interpreted to "the operating environment".

Furthermore, part 2(a) of this section requires enterprises to provide the public with measurable, verifiable, and timely information on the potential environmental and safety impacts of their activities.

---

[16] The Ten Principles of the UN Global Compact, www.unglobalcompact.org/what-is-gc/mission/principles (2011).

[17] OECD Declaration and Decisions on International Investment and Multinational Enterprises, www.oecd.org/investment/investment-policy/oecddeclarationanddecisions.htm (2012).

[18] Morgera, Elisa, OECD Guidelines for Multinational Enterprises, In *Handbook of Transnational Governance*, Polity, www.oecd.org/corporate/mne/ (2011): 314–322.

Guideline 6(d) of the environment section also asks the enterprise to explore methods for improving their environmental performance in different ways such as efficient use of the resources.[19]

On the other hand, although multinational enterprises are defined as business entities that have their home in one country but operate in other countries as well,[20] it can be interpreted that since the space companies conduct their operations in the outer space where is a region beyond the jurisdiction of any country including the home country of the space company, private space enterprises can also be regarded as multinational or transnational enterprises and can be included in the scope of the OECD guidelines. In this way, the useful guidelines of OECD, especially in the environment section, can be applied for the space companies conducting the business of mega-constellations of satellites.

3 The Capacity of Space Law to Incorporate Environmental Principles of CSR into Satellite Mega-Constellations Projects

In the following section, it is argued that the principles of CSR are in line with the current regime of international space law by examining the most relevant international space regulations, and that the implementation of international principles and rules in outer space will certainly help to maintain the sustainability of outer space.[21] It is presumed that the CSR principles are not changing the content of the current legal regime of outer space but rather tries to normalize mega-constellation activities and adapt them into international space law. As will be shown the volume of space treaties and resolutions alongside the customary international law could help CSR to apply smoothly.

3.1 Relevant Hard Law Provisions

Generally, CSR is trying to regulate the private sector activities in a more responsible way. The legal regime of outer space has evolved through the adoption of both resolutions and treaties which is accompanied by customary international law. These documents mostly are the result of the efforts of the Committee on the Peaceful

---

[19] Ibid. *supra note* 19.

[20] S. H. Mahmoudi, M. Sedighian Kashani, Human Rights Related Obligations and Transnational Corporations with Emphasis on the Approach of the Human Rights Council, Journal of Comparative Law Studies, Vol. 9, No. 2 (2018):829–849.

[21] T. Lundsgaard, CSR in Space: Corporate Social Responsibility Principles for the Space Industries, SSRN 2687703 (2015).

Uses of Outer Space. What is considerable in these documents is the dominant presence of States in outer space. However international space law does not prevent the involvement of non-governmental space activities. In practice, privatization and commercialization of space activities began during the 1980s and in particular 1990s in the United States and in some space-faring States.[22] There is considerable evidence in space treaties showing that private activities in outer space are permitted. Private space activities are governed by both national and international law. Among the space activities, the mega satellite constellations in LEO are going to be operated by the private sectors. Precisely because CSR seeks to encourage commercial companies to behave more responsibly, these principles can run parallel to national and international space law regulations for companies operating this system. CSR principles could compel the mega satellite constellations companies to be in conformity with environmental and human rights dimensions and be in the service of the sustainability of space activities. Considering the various.

Given the vast spatial and operational dimensions of satellite systems, the need to apply CSR to such activities is essential. The number of active satellites in a system is very large. They are interconnected in the form of a network. A large area of space is occupied by such satellites. These satellites are basically small satellites and probably have a shorter lifespan, resulting in increased waste in LEO.[23] The scope of these activities inevitably limits the activities of other countries, especially developing countries. Possible concerns about the threats these systems pose to the space environment require that, in addition to national and international regulations, they use CSR principles to control the behavior of these mega satellite constellations companies. In addition, the services of these satellites are directly related to the daily lives of the people, and their improper operation can be a direct threat to the international community.

According to Article I of the Outer Space Treaty "The exploration and use of outer space, including the Moon and other celestial bodies, shall be carried out for the benefit and in the interests of all countries, irrespective of their degree of economic or scientific development, and shall be the province of all mankind".[24]

In this regard, the question is how to ensure that a satellite constellation company implements this Article well. Naturally, the Outer Space Treaty commits governments, but at the same time indirectly controls the behavior of private companies. It also manages member states. There is no doubt that the services of satellite systems are for the benefit of humanity, but the question is whether companies operating in this field, given their social responsibility, ensure that the interests of countries are

---

[22] Ibid. *supra note 2*.

[23] M. Haddadi, S. H. Mahmoudi, R. Madadi, Liability for Damage caused by Launching Small Satellites by Space Companies and Startups, *Private Law* (2020) (in Persian).

[24] Treaty on Principles Governing the Activities of States in the Exploration and Use of Outer Space, including the Moon and Other Celestial Bodies (adopted 27 January 1967, entered into force 10 October 1967) (Outer Space Treaty) art I.

considered regardless of their degree of scientific, political, and economic development. It can be argued that compliance with the CSR principles effectively contributes to the implementation of Article I of the Outer Space Treaty.

Moreover, Article III of Outer Space treaty stipulates that "States Parties to the Treaty shall carry on activities in the exploration and use of outer space, including the Moon and other celestial bodies, in accordance with international law, including the Charter of the United Nations, in the interest of maintaining international peace and security and promoting international cooperation and understanding".[25]

This article shows that international space law is a part of the system of international law. This article, therefore, paves the way for the application of other international regulations in the field of space. CSR principles can be part of this effort. Another is that CSR can undoubtedly prevent the temptation of satellite constellation companies to use their services in opposition to international peace and security. This becomes even more important when the State(s) of the registry of constellations enters into an international conflict with another State and forces those companies to be in the service of that conflict.

Despite the fact that the Outer Space Treaty is designed in the atmosphere of interstate space activities, it should not be overlooked that Article VI of OST has regulated the relationship between non-governmental space activities and the relevant State party. Contrary to the general rules governing the international responsibility of states, whose current reflection can be seen in the Draft Articles of International Responsibility of States for Internationally Wrongful Acts,[26] and accordingly, in principle, states have no international responsibility for the actions of private individuals, we consider that the international law of space has chosen the opposite trend. This attitude is quite clear in Article VI of the OST in accordance with this article:

> States Parties to the Treaty shall bear international responsibility for national activities in outer space, including the Moon and other celestial bodies, whether such activities are carried on by governmental agencies or by non-governmental entities, and for assuring that national activities are carried out in conformity with the provisions set forth in the present Treaty. The activities of non-governmental entities in outer space, including the Moon and other celestial bodies, shall require authorization and continuing supervision by the appropriate State Party to the Treaty. When activities are carried on in outer space, including the Moon and other celestial bodies, by an international organization, responsibility for compliance with this Treaty shall be borne both by the international organization and by the States Parties to the Treaty participating in such organization.[27]

Notably, this article is pertaining to the responsibility of State parties for the national activities whether such activities are carried on by governmental agents or by non-governmental entities. The important point in this article is the obligation of States for authorization and continuing supervision of the activities of the non-government entities.

---

[25] Ibid. *supra note* 24, art III.

[26] Responsibility of States for Internationally Wrongful Acts, with commentaries, https://legal.un.org/ilc/texts/instruments/english/commentaries/9_6_2001.pdf (2001).

[27] Ibid. *supra note* 24, art VI.

Normally the authorization and continuing supervision are carried on through the national space legislation and policy. In this regard, it is important that the authorization and continuing supervision extend to the concept of CSR principles. Indeed, the principles of CSR are not legally binding, but at least at the national level, CSR can be incorporated into the national legal system of States. This can be done through the inclusion in the National Space Law, or the principles of CSR can be independently contained in a separate document, and mechanisms for compliance by companies can be established. It is a useful tool for States to force mega satellite companies to act in conformity with the environmental aspects of international space law.

The most relevant provision relating to the environment is Article IX of the Outer Space Treaty. This article introduces several important provisions. First, it reminds that in space activities, they should be guided by the principle of cooperation and mutual assistance, and in their activities, they should observe the interests with due care. Another related obligation is the commitment to avoid harmful contamination. This commitment is precisely covered by the environmental principles of CSR. Accordingly, under Article IX: "States Parties to the Treaty shall pursue studies of outer space, including the Moon and other celestial bodies, and conduct exploration of them to avoid their harmful contamination and also adverse changes in the environment of the Earth resulting from the introduction of extraterrestrial matter and, where necessary, shall adopt appropriate measures for this purpose."[28]

Specifically, one of the appropriate measures to prevent harmful contamination is to require satellite companies to follow the principles of CSR. This Article adds that:

> If a State Party to the Treaty has reason to believe that an activity or experiment planned by it or its nationals in outer space, including the Moon and other celestial bodies, would cause potentially harmful interference with activities of other States Parties in the peaceful exploration and use of outer space, including the Moon and other celestial bodies, it shall undertake appropriate international consultations before proceeding with any such activity or experiment. A State Party to the Treaty which has reason to believe that an activity or experiment planned by another State Party in outer space, including the Moon and other celestial bodies, would cause potentially harmful interference with activities in the peaceful exploration and use of outer space, including the Moon and other celestial bodies, may request consultation concerning the activity or experiment.[29]

In general, the satellite system project is very large and extensive. Therefore, the probability of risk is very high, and the provisions of Article 9 should undoubtedly be considered in the operation of satellite systems. In this regard, the principles of CSR can be areas for facilitating international consultation.

---

[28] Ibid. *supra note* 24, art IX.

[29] Ibid. *supra note* 24, art IX.

3.2 Relevant Soft Law Instruments

Among the UN space resolutions, the "Principles Relevant to the Use of Nuclear Power Sources in Outer Space".[30] Resolution is relevant to the environmental aspects of mega satellite constellations. What can be said about the proper implementation of this resolution is that it has imposed restrictions on the use of nuclear power sources for satellites. In general, given that the satellite constellations are normally in LEO orbit, the use of nuclear power for a constellation is hardly in conformity with the aforementioned resolution.

Another important resolution in this regard is the "Declaration on International Cooperation in the Exploration and Use of Outer Space for the Benefit and in the Interest of All States, Taking into Particular Account the Needs of Developing Countries". This resolution is in fact a re-emphasis on some of the principles set out in the OST, including the principle of international cooperation and freedom of use and exploration and reaffirms that exploration and use of outer space "shall be carried out for the benefit and in the interest of all States, irrespective of their degree of economic, social or scientific and technological development, and shall be the province of all mankind. Particular account should be taken of the needs of developing countries".[31] Another important provision of this resolution that is very significant for the operation of mega satellite constellations is as follows:

> International cooperation should be conducted in the modes that are considered most effective and appropriate by the countries concerned, including, inter alia, governmental and non-governmental; commercial and non-commercial; global, multilateral, regional or bilateral; and international cooperation among countries in all levels of development." In line with the implementation of this clause, it can be noted that adherence to the principles of CSR by satellite constellations companies is a clear example of "international cooperation between countries at all levels of development.[32]
>
> In line with the implementation of this clause, it can be noted that adherence to the principles of CSR by satellite constellations companies is a clear example of "international cooperation between countries at all levels of development.

Moreover, the international guidelines regarding the space debris mitigation issue should be taken into consideration. Among the relevant international guidelines, the Inter-Agency Space Debris Coordination Committee (IADC) guidelines on Space Debris Mitigation[33] were adopted in 2002 and revised in 2007 and UNCOPUOS

---

[30] United Nations Resolution 47/68, Principles Relevant to the Use of Nuclear Power Sources in Outer Space, www.unoosa.org/oosa/en/ourwork/spacelaw/principles/nps-principles.html.

[31] United Nations Resolution 51/122, Declaration on International Cooperation in the Exploration and Use of Outer Space for the Benefit and in the Interest of All States, Taking into Particular Account the Needs of Developing Countries, www.unoosa.org/oosa/en/ourwork/spacelaw/principles/space-benefits-declaration.html.

[32] Ibid. *supra note* 31.

[33] Inter-Agency Space Debris Coordination Committee, Inter-Agency Space Debris Coordination Committee Home Page, www.iadc-online.org/.

Space Debris Mitigation Guidelines[34] adopted in 2007, are of the most relevant and significance. Since the UNCOPUOS guidelines are based on the IADC document, they are mostly similar. The seven main guidelines of these guidelines require States to:

1. Limit debris released during normal operations
2. Minimize the potential for break-ups during operational phases
3. Limit the probability of accidental collision in orbit
4. Avoid intentional destruction and other harmful activities
5. Minimize the potential for post-mission break-ups resulting from stored energy
6. Limit the long-term presence of spacecraft and launch vehicle orbital stages in the Low-Earth orbit (LEO) region after the end of their mission
7. Limit the long-term interference of spacecraft and launch vehicle orbital stages with the geosynchronous Earth orbit (GEO) region after the end of their mission.

In 2018 the Guidelines for Long Term Sustainability of Outer Space Activities (LTSSA) were adopted, guiding the policy and regulatory framework for space activities in member states. As well as mitigation guidelines, the subcommittee also encourages the member states to ensure the implementation of LTSSA guidelines voluntarily. Many parts of the LTSSA guidelines are relevant to environmental principles of CSR such as:

> Guideline B.3: According to these guidelines, states are encouraged to develop and use the technologies related to monitoring space debris and share the data for research and scientific cooperation with other states.
>
> Guideline B.8.2: The manufacturers and operators of space objects should be encouraged by their state to design the object with regard to space debris mitigation guidelines or standards.
>
> Guideline C.4.4: The cooperation between states and its non-governmental entities should be fostered since such entities play an important role in the increase of awareness of issues related to space sustainability such as debris mitigation guidelines, among industry associations and academic institutions.[35]

Last but not least, the space debris mitigation standards of the International Organization for Standardization (ISO)[36] published in 2010, in its latest version which is 24113:2019 space systems—space debris mitigation requirements, provides a set of engineering space debris mitigation standards.

All these environmental provisions of space law prepare the ground for integrating the environmental principles of CSR into space law. These two sets of frameworks can

---

[34] United Nations, Office For and Outer Space, 'United Nations Space Debris Mitigation Guidelines of the Committee on the Peaceful Uses of Outer Space'.

[35] Long-term Sustainability, Outer Space Activities and I Context, 'Committee on the Peaceful Uses of Outer Space Guidelines for the Long-Term Sustainability of Outer Space Activities Conference Room Paper by the Chair of the Working Group on the Long-Term Sustainability of Outer Space Activities I. Context of the Guidelines for the Long-Term Sustainability of Outer Space Activities' (2018) 04441.

[36] International Standards and others, 'ISO 24113: 2019 Space Systems—Space Debris Mitigation Requirements'.

interact with and complement each other. More specifically, environmental principles of CSR can help to maintain the sustainability of outer space activities.

4 Application of Environmental Principles of CSR to Adverse Impacts of Satellite Mega-Constellations

Mega-constellations of satellites are the main topic of discussion in today's technical and legal communities of the space sector. The primary objective of the manufacturing companies of recent large constellations is to provide affordable, global, and high-speed internet access even for areas where no internet access is available or it is limited. Even though this goal could be regarded as a philanthropic goal, especially for deprived areas, its potential adverse impacts are undeniable. The two main impacts that mega-constellations might have are the extreme effect on the astronomy and observation of the night sky[37] and the conundrum of space debris generation and high risk of orbital collisions. While the first is one of the main topics of discussion of scientists and astronomers at the international level, it is more relevant to the social impacts of mega-constellations projects as a business. Therefore, here the environmental impacts of such projects i.e., space debris will be examined and it is discussed that companies of mega-constellations of satellites, as business corporations, should comply with one of the most central principles of CSR, which is preventing businesses from posing adverse impacts on the environment.

4.1 Overview of Two Main Mega-Constellations Projects

The two main mega-constellations projects are proposed and being launched by the SpaceX and OneWeb companies. SpaceX's constellation called Starlink is a satellite internet constellation which has plans to deploy 30,000 satellites in LEO (at the altitude of around 600 km). On Feb the first, 2019 Space X filled the application for the U.S. Federal Communication Commission (hereinafter FCC) license for earth station services. In its application, SpaceX mentions that "Granting this application would serve the public interest by helping to speed broadband deployment throughout the United States by authorizing the ground-based component of SpaceX's satellite system."[38]

FCC granted a license to Starlink "Blanket Licensed Earth Stations" application on 18 March 2020 for 15 years accepting that granting the application will sever the public interest. Since May 2019, SpaceX has been launching its new Starlink internet

---

[37] Report of UNOOSA and IAU workshop on Dark and Quiet Skies for Science and Society, www.iau.org/static/publications/dqskies-book-29-12-20.pdf (2020).

[38] FCC, SpaceX Application for blanket licensed earth stations, https://fcc.report/IBFS/SES-LIC-INTR2019-00217/1616678.pdf (2019).

satellites in batches of 60. The latest launch of Starlink satellite was on 20 January 2021 and with this launch the total number of Starlink constellation which have been so far launched reached 1015.[39]

The other mega satellite constellation project is the OneWeb satellite constellation which like Starlink aims to provide global satellite internet. OneWeb initially planned to deploy 650 satellites in LEO (at the altitude of 1200 km) the plan which changed later. Despite the fact that in March 2020, OneWeb announced its bankruptcy,[40] it filed an application to FCC and proposed to increase its current number of satellites to 48,000. OneWeb launched 74 of its satellites before its bankruptcy. Later in November 2020, OneWeb could exit bankruptcy and recently on 12 January 2021, requested FCC to modify the number of proposed satellites in its previous application from 48,000 to 6372.[41]

4.2 CSR Principles and Mega-Constellations' Space Debris

Large satellite constellations do not pose any threat to the space debris problem, per se. The main threat that such constellations of satellites cause are the augmentation of the collision possibility rate and the creation of space debris.[42] As mentioned above, currently the most popular orbit for the deployment of satellite constellations is the LEO due to reasons discussed earlier. Plus, in the first chapter, it is explained that LEO orbits are the most crowded Earth orbits with a high density of active and inactive satellites and spacecraft. Thus, constellations launched and deployed in LEO are more hazardous to the space debris environment since they will increase the density of spacecraft in such orbits even more. This will lead to higher collision risk possibilities. If two space objects in this orbit collide, there is the chance that the collision will be continued like a chain among other objects in orbit creating a catastrophic situation. It was shown in a study in 2016 that if the post-mission disposal (PMD) time of satellite constellations decrease and the success rate of PMD increases to 90%, the negative impact of the large constellations on the debris environment in the long and short term will also decrease.[43] The debris impact of satellite constellations, however, is not limited to the post-mission time. Debris can also be created in the

---

[39] SpaceX surpasses 1000-satellite mark in latest Starlink launch, https://spacenews.com/spacex-surpasses-1000-satellite-mark-in-latest-starlink-launch/ (2021).

[40] Bankruptcy court approves OneWeb sale, https://spacenews.com/bankruptcy-court-approves-oneweb-sale/ (2020).

[41] OneWeb slashes size of future satellite constellation, https://spacenews.com/oneweb-slashes-size-offuture-satelliteconstellation/#:~:text=WASHINGTON%20%E2%80%94%20OneWeb%20says%20it's%20drastically,to%20have%20nearly%2048%2C000%20satellites.&text=The%20satellites%20in%20the%2087.9,the%20system%20to%206%2C372%20satellites (2021).

[42] Jakhu Ram S., and Joseph N. Pelton, eds, *Global space governance: an international study*, Springer International Publishing (2017): 363.

[43] C. Chen, Y. Wulin, The impact of large constellations on space debris environment and its Countermeasures (2017).

launch, deployment, and operation phases of large constellations. Another research has been conducted by the French National Centre for Space Studies (CNES) and the U.S. National Aeronautics and Space Administration (NASA) using different debris models for examining the OneWeb project (including 720 satellite) as an example to show the impact risk of large constellations during launch and operation and PMD phases. According to the results of their studies, the higher impact risk during the lifetime of a large constellation is dedicated to the debris created in the PMD phase. The launch phase, however, has the lowest probability of collision risk since the duration of deployment is short. Moreover, the orbit height of the OneWeb satellites is 1200 km above the Earth which is not a densely populated orbit making the debris impact risk in the operation phase low as well. The figure below shows the results of this research.

The study later analyzes OneWeb constellation from the perspective of debris impact of large satellite constellations in three different stages of their life cycle on the existing space objects in orbit. The result is that in the launch and deployment stage, the risk of intersection between the satellites of the constellation and the existing space object in the target orbit is low due to the short period of launch. During the operation stage, although the destination orbit of the OneWeb constellation is not overcrowded, the risk of rendezvous is higher since the density of the constellation satellites is high. In the PDM stage, the number of intersections is considered to be the highest because the time of the PDM stage in the OneWeb project is long (5 years after the end of the mission) and also a lot of space objects pass through LEO orbit. However, the study shows that if the period of PMD decreases to 3 or 1 year, the intersection rate with other space objects will consequently decrease.

It can be concluded that the collision rate in LEO will dramatically increase by the installation of large satellite constellations in this area. This becomes more significant when it comes to the International Space Station (ISS) The ISS is also located in LEO which means the increase in the impact debris probability of constellations will not only pose a great danger to the station itself but also the life of the astronauts permanently working in the station.[44]

According to principle 7 of UNGC, businesses are required to undertake precautionary approaches to environmental challenges. There is no doubt in the fact that space debris is the most dangerous challenge for the space environment and the future activities in outer space. Also, based on the above-mentioned arguments and statistics, mega-constellations of satellites pose a serious threat to the orbital environment where they operate and can result in potential consequent risks such as cascade collisions and even falling the remaining parts of debris on the Earth! From the perspective of space debris generation, one can hardly conclude that the ongoing mega-constellations business is of a responsible and sustainable kind and considers the relevant environmental principles of CSR.

It is also worth to mention that the obligation of businesses to develop and diffuse environmentally friendly technologies outlined in principle 9 of the UNGC should

---

[44] A. Barabi, X. Chunli, Legal Aspects of Space Debris: Implication of Large Satellite Constellations, master's dissertation, Beihang University (2020).

be taken in account by the operators of mega constellations of satellites in order to prevent and minimize generation of space debris. An example of environmentally friendly space technology could be the recent "wooden satellite" proposed by a Japanese company and Kyoto University. On December 2020, Sumitomo Forestry announced that they are planning to launch the "world's first satellite made out of wood" in 2023 because wooden satellites will be fully burnt in the atmosphere and subsequently no debris will be left in space or fall on the ground. For this purpose, they are experimenting different types of woods resistant to temperature changes and sunlight.[45]

5 Conclusion

Projects of mega satellite constellations are being formed and it is predicted that in the near future, these satellites will occupy the entire orbit of LEO. Without ignoring the benefits of these constellations, their negative effects and results should not be neglected either. The current legal regime of outer space also has regulations on the space environment. Nevertheless, in addition to the applicable regulations in outer space, the non-binding but important principles of CSR can serve as a complementary resource to the long-term sustainability goals of space activities. In fact, space law regulations and CSR principles in the field of the environment have a common concern but different forms. While space law regulations typically govern States' obligations, the CSR aims to make companies directly and socially accountable for risks and threats.

Finally, it is suggested that the CSR principles be applied in two forms for satellite constellations: First, these companies should be unilaterally committed to these principles and try to contribute to their efficiency by observing these principles in an environment where the possibility of competition between them increases. By observing these principles, they can also contribute to their efficiency. Second, it is proposed that the principles of CSR be enshrined in the national space laws of States, and that authorization for such activities should be subject to compliance with these principles. In general, satellite constellations are subject to the international space law. However, since these projects are usually operated by commercial companies, they must comply with the requirements of Article VI of the Outer Space. Although under Article VI OST, States are responsible for the activities of non-governmental entities, they may extend the authorization and continuing supervision set forth in Article VI to the principles of CSR. It seems that if there is competition for such activities, the probability of following these principles increases.

---

[45] Japan developing wooden satellites to cut space junk, www.bbc.com/news/business-55463366 (2020).

S. Hadi Mahmoudi is an Assistant Professor at the Department of International Law, Faculty of Law, Shahid Beheshti University, Tehran, Iran. His focus is on international air and space law and has published several works on international law. He is a member of the International Institute of Space Law (IISL) as well as the Representative of the Ministry of Science, Research, and Technology in the Subcommittee on Space Law and Policy of Supreme Space Council, Iran.

Aishin Barabi is a Legal Researcher at the Responsible Business Promotion Institute, Tehran, Iran. She has gained her masters' degree in Space Law and Policy major from Beihang University, Beijing, China. Her thesis topic was "Legal Aspects of Space Debris: Implication for Large Satellite Constellations". She has also graduated with a bachelor's degree in Law from Shahid Beheshti University, Tehran.

International Cooperation as an Essential Part of the Galileo Programme

Tugrul Cakir

Abstract The issue of compatibility and interoperability of navigation systems requires cooperation among service operators. In addition, cooperation as integral part of space industrial policy is a powerful enabler to enhancing competitiveness of the space industry. The purpose of this chapter is to analyze how international cooperation is developed and implemented by the European Union under the Galileo Programme. This chapter argues that international cooperation is an important tool in service of governance.

1 Introduction

The context of space has radically changed with the intensification and diversification of space activities. Now, modern society has become increasingly dependent on these activities. The disruption of these activities, even if only for a limited time, has the very real potential to paralyze our daily lives. Moreover, the use of space technologies in the management of the Covid-19 outbreak, highlighted the crucial role played by these technologies at a critically important time.[1]

The main functions of satellite navigation systems are positioning, navigation and precise timing. They make our everyday lives easier, and are used in a great number of fields of activity, such as search and rescue, road traffic management, air navigation, financial operations, border control, precision agriculture and natural resources management etc. As of December 2019, four Global Navigation Satellite

---

[1] Tanja Masson-Zwaan, "Combating COVID-19: The Role of Space Law and Technology" (2020) 45 Air & Space Law 39–60; ESPI Special Report, "COVID-19 and the European space sector" (2020) 16.

T. Cakir (✉)
Faculty of Law, Ankara Yildirim Beyazit University, Ankara, Turkey
e-mail: tcakir@ybu.edu.tr

© The Author(s), under exclusive license to Springer Nature Switzerland AG 2021
A. Froehlich (ed.), *Legal Aspects Around Satellite Constellations*,
Studies in Space Policy 31, https://doi.org/10.1007/978-3-030-71385-0_8

Table 1 Table of operational PNT satellites[a]

Constellation-Country/International organization	Number of operational satellites/Total satellites in constellation	Share of global operational satellites (%)
BeiDou-China	43/48	32.8
GPS-United States of America	31/31	23.6
GLONASS-Russian Federation	24/27	18.3
Galileo-European Union	22/26	16.7
NavIC-India (regional)	7/7	5.3
QZSS-Japan (regional)	4/4	3
Total	131/143	

[a]Space Foundation, The Space Report (2020) Q2 44–47

System (GNSS) and two Regional Navigation Satellite System (RNSS) are operational.[2] Table 1 shows the share of the operational PNT (Positioning, Navigation and Timing) satellites at the end of 2019.

The Galileo Programme is the first GNSS conceived for civilian needs and under civilian control as opposed to the GPS, GLONASS and BeiDou designed for military purposes. Galileo is a joint project of the European Union (EU), the European Space Agency (ESA) and the European space industry.[3] The programme is funded and owned by the EU and the European Commission has the overall responsibility for Galileo. The ESA acts as the design and procurement agent on the behalf of the Commission. The European Global Navigation Satellite Agency (GSA),[4] a regulatory agency of the EU, has been entrusted with the operational management of the programme since the formal handover (from ESA) in July 2017. Galileo Initial Services (which are the Open Service (OS), the Search and Rescue support

---

[2]In addition to these, there are also satellite navigation augmentation systems (Ground Based Augmentation Systems (GBAS) and Satellite Based Augmentation Systems (SBAS)) developed in order to enhance the accuracy of navigation satellites. See generally on these systems Olga Volynskaya, "Navigation by satellite", in Philippe Achilleas, Stephan Hobe (eds) Fifty Years of Space Law (Brill 2020) 400–406. The main SBAS networks include: American Wide Area Augmentation System WAAS (USA), European Geostationary Navigation Overlay Service EGNOS (EU), Multi-functional Transport Satellite MSAS (Japan), GPS and GEO-augmented Navigation GAGAN (India), System of Differential Correction and Monitoring SDCM (Russian Federation).

[3]On the European integration process in the space field see generally Nina-Louisa Remuss, Theorising Institutional Change: The Impact of the European Integration Process on the Development of Space Activities in Europe (Springer 2018). The author focuses in particular on examining the impact of the European integration process on other actors and offers innovative avenues on the future of the governance of space activities in Europe. On the issue of Galileo governance see generally Amiel Sitruk, Serge Plattard, "The governance of Galileo" (2017) ESPI Report 62.

[4]The GSA will be transformed into the European Union Space Programme Agency (EUSPA) in early 2021. GSA, "End-of-year message from GSA executive director Rodriga da Costa", Published 23 December 2020, www.gsa.europa.eu/newsroom/news/end-year-message-gsa-executive-director-rodrigo-da-costa All websites cited in this publication were last accessed and verified on 9 January 2021.

Service (SAR), and the Public Regulated Service (PRS) for pilot tests) commenced in December 2016. Galileo will become fully operational in early 2021 and the full operational constellation will consist of 24 satellites and 6 spares.

Strengthening Europe's role as a global actor and the promotion of international cooperation is one of the pillars of the European Space Strategy according to the Space Strategy of European Commission, announced in October 2016.[5] The promotion of cooperation under the Galileo Programme is also an integral part of this strategy. It is the first time, since the end of the 1990s, that in global navigation, an "active international co-operation" has been implemented from the beginning of a programme.[6]

Indeed, international cooperation is necessary in the Multi-GNSS world. The main rationale of international cooperation stems from both compatibility and interoperability concerns with the other GNSS systems. The former is defined as "the ability of GNSS and regional systems to be used separately or together without causing interference[7] with each other's individual service or signals, and without adversely affecting national security".[8] The latter is characterized as "the ability of global and regional satellite systems, augmentations and services, to be used together seamlessly to provide better capabilities at user level than would be achieved by individual systems".[9] The main benefits of interoperability are better availability and accuracy of systems. Even if, a multiplicity of GNSS can be seen unnecessary as an initial reaction, this multiplicity offers a way to compensate for the "inherently fragile" character of GNSS.[10] Interoperability is also linked to the downstream market which has become increasingly competitive. Table 2 shows GNSS downstream market revenues from devices and services in 2019 and expected in 2029.

The second rationale for cooperation in the GNSS field is to allow market access and to maintain and foster competitiveness of the space industry at global level. Developing and implementing industrial cooperation allows a service provider to encourage economic growth by introducing and supporting the use of its GNSS

---

[5] European Commission, "Space strategy for Europe", Brussels, 26 October 2016, COM (2016) 705 final, 11–12, https://ec.europa.eu/docsroom/documents/19442. The other pillars are maximizing the benefits of space for society and the EU economy, fostering a globally competitive and innovative European space sector, reinforcing Europe's autonomy in accessing and using space in a secure and safe environment.

[6] Eero Ailio, "International co-operation in Galileo", Aurélien Desingly (ed), *Galileo, la navigation par satellite européenne, questions juridiques et politiques au temps de la concession* (IFRI 2006) 206.

[7] Harmful interference may be defined as "an interference which endangers the functioning of a radionavigation service or of other safety services or seriously degrades, obstructs, or repeatedly interrupts a radiocommunication service operating in accordance with Radio Regulations". Article 1.169 of ITU Radio Regulations, Edition of 2020.

[8] UNOOSA, "International Committee on Global Navigation Satellite Systems, The way forward, 10 years of achievement 2005–2015" (2016) 31.

[9] Ibid. On the issue of interoperability see generally Günter Hein, "GNSS Interoperability: Achieving a Global System of Systems or 'Does Everything Have to be the Same?'" (2006) Premiere Issue Inside GNSS 57–60.

[10] Francis Lyall, Paul B. Larsen, *Space Law A Treatise* (Ashgate 2009) 401.

Table 2 GNSS downstream market revenues and expected revenues[a]

	2019 revenue (€)	Share of world (%)	2029 revenue (€)	Share of world (%)
EU (EU 28)	38.4 billion	25.5	65.3 billion	20.1
North America	40.3 billion	26.7	92.2 billion	28.4
Asia Pacifique	46.0 billion	30.5	106.0 billion	32.7
Other Regions	26.0 billion	17.3	60.9 billion	18.8
World	150.7 billion		324.4 billion	

[a]GSA, "GNSS Market Report-Issue 6" (2019) 6

outside its territory (or region). In so doing, the national (or regional) space industry is able to take advantage of business opportunities especially in lucrative markets. More generally, space industrial cooperation is an important tool to strengthen the industry and to promote the use of its GNSS technologies and services in the face of increasing competition in the market.

Before delving into the question of cooperation in the field of satellite navigation (3), and into the manner in which the international cooperation is developed and implemented under the Galileo Programme (4), it is worthwhile reviewing the main features of the services provided by Galileo (2).

2 Galileo: The First Flagship Programme[11] of the European Union

The European Union has, for a long time, had an increasing interest in space related fields even more after a *sui generis* shared competence, between the European Union and the Member States, provided by the Lisbon Treaty (Treaty on the Functioning of the European Union (TFEU) sealed 18 December 2007 and effective 1 December 2009).[12] Before that time, no specific competence had been attributed by a text to the European institutions in the space field. This competence allows the Union to develop a common European space policy and to establish a space programme. With this Treaty, the EU's role in space has been reinforced, which allows it to be an

---

[11]"In a generic sense, flagship programmes address grand scientific and societal challenges, which require a common European research effort and sustained support for a development period of at least ten years. They represent science-driven, large-scale, multidisciplinary research initiatives oriented towards a unifying goal, which is expected to have a transformational impact on science and technology and substantial benefits for European competitiveness and society. The goals of such initiatives are visionary and highly ambitious in terms of scientific chal- lenges, resources required and coordination of efforts. In terms of implementation and operations, they require cooperation between a range of disciplines, communities and programmes, including national and European initiatives." Marco Aliberti, Arne Lahcen, "The Future of European Flagship Programmes in Space" (2015) 53 ESPI Report 39.

[12]See generally Frans von der Dunk, "The EU space competence as per the treaty of Lisbon: sea change or empty shell?" (2011) 61 Proc. on L. Outer Space 382–392.

important player not only in Europe, but also in the international arena. The draft International Code of Conduct for Outer Space Activities is one of the first important steps in the creation of a new main actor in the space field.[13]

The legal basis of the development of the Galileo Programme is Article 170 of the TFEU (ex-Article 154 TEC), which provides that "the Union shall contribute to the establishment and development of trans-European networks in the areas of transport, telecommunications and energy infrastructures". In Europe's perspective, space was originally perceived and pursued for social and economic benefits.[14] This is also the case for the Galileo Programme. It does not mean that this programme has not also been designed to enhance Europe's security.[15] It allows Europe to have its own independent system controlled by the EU and its Member States.[16] Therefore, it can be considered "a major sovereignty tool".[17] This strategic autonomy is beneficial for Europe from an economical and security point of view in a growing number of sectors.[18] More generally, Galileo is crucial in the development, implementation and monitoring of the EU's and Member State's policies.[19]

The initial services offered by Galileo are:

Open Service (OS): A free of charge OS (with a provision of basic signals) has been offered to everyone throughout the world, to everyone who has the appropriate receiver. There is no contractual guarantee as to its continuity and integrity. In the near future, Navigation Message Authentication (Open Service-NMA), provided on E1 and E5 signals, "will allow users to verify that a navigation message comes from a Galileo satellite and not a potentially malicious source".[20] This additional feature of the OS, which will highly be important for protection against spoofing attacks, will be offered free of charge.[21]

---

[13] Even if the project had failed to achieve consensus at the 2015 international conference, the experience gained during the development process of this Code might be beneficial for a future space traffic management regime. Kai-Uwe Schrogl, "Space Law and diplomacy" (2017) 67 Proc. on L. Outer Space 5–6.

[14] Vladimír Remek, Christina Giannopapa, "Perspectives on Anchoring the Utilisation of EGNOS/Galileo and GMES and Other Space Applications" (2011) 49 ESPI Perspectives 6. See in this sense Council of the European Union, "Space as an enabler" (28 May 2019) para.1 www.consilium.europa.eu/media/39527/20190528-council-conclusions-space.pdf.

[15] On the dual use approach of the EU see generally Nunzia Paradiso, "The EU dual approach to security and space, Twenty Years of European Policy Making" (2013) 45 ESPI Report.

[16] Christina Giannopapa, "The Less Known, but Crucial Elements of the European Space Flagship Programmes: Public Perception and International Aspects of Galileo/EGNOS and GMES" (2011) 34 ESPI Report 6.

[17] Sitruk, Plattard (no 3) 5.

[18] Daniel Fiott, "The European space sector as an enabler of EU strategic autonomy", In-depth analysis requested by the SEDE Subcommittee, Policy Department for External Relations Directorate General for External Policies of the Union PE 653.620 (December 2020) 6. www.europarl.europa.eu/thinktank/en/document.html?reference=EXPO_IDA(2020)653620.

[19] Giannopapa (no 16) 34.

[20] GSA, "GNSS Market Report-Issue 6" (2019) 15.

[21] Ibid.

Search and Rescue Service (SAR): This service contributes to the service provided by the COSPAS-SARSAT system. After detecting emergency signals from individuals, ships and aeroplanes in distress, this service relays them to Rescue Coordination Centers. This service is available worldwide and provided without a fee. Since January 2020, Return Link Service (RLS), designed by the ESA, has been improving the service such that, in a couple of minutes, it is able to assure the users in distress, that help is on the way.[22] Galileo is the first GNSS providing this capability.

Public Regulated Service (PRS): This service is used in sensitive applications such as police, ambulance services, fire brigades and coast guards etc. PRS is also vital for military uses of Galileo and responds to the security requirements.[23] These signals are highly encrypted and resistant to jamming and spoofing. Member States, the Council, the Commission and the European External Action Service (EEAS) have the right to access to the PRS.[24] The decision as to whether or not to use the PRS signals, is left to the Member States to decide.[25] International organizations and third countries may become PRS participants by way of separate agreements concluded with the Union.[26] Such agreements could cover the manufacturing, under specific conditions, of PRS receivers, with the exclusion of security modules.[27] European values such as democracy and the rule of law should be taken into account during the negotiation process of these agreements.[28] The US requested access to PRS signals and negotiations are ongoing.[29]

Once fully operational, Galileo will offer:

High Accuracy Service (HAS): This service will "complement the OS, delivering high accuracy data and providing better ranging accuracy, enabling users to achieve sub-meter level positioning accuracy".[30] It will be provided without a fee.

Commercial Authentication Service (CAS): Contrary to the other GNSS, Galileo will offer a commercial service. This service will "complement the OS, providing a

---

[22] Galileo GNSS, "Galileo now replying to SOS messages Worldwide" (Galileo GNSS, 10 February 2020) https://galileognss.eu/galileo-now-replying-to-sos-messages-worldwide/.

[23] Sitruk, Plattard (no 3) 59.

[24] Decision no 1104/2011/EU of the European Parliament and of the Council of 25 October 2011 on the rules for access to the Public Regulated Service provided by the global navigation satellite system established under the Galileo programme, Article 3.1. www.gsa.europa.eu/sites/default/files/decision_no_1104-2011-eu_on_the_rules_for_access_to_the_public_regulated_service_provided_by_the_global_navigation_satellite.pdf.

[25] Ibid., Article 3.2.

[26] Ibid., Article 3.5.

[27] Ibid., Article 3.5(b).

[28] Ibid., Preamble para.9.

[29] ESPI, "Security in Outer Space: Perspectives on Transatlantic Relations" (2018) 66 ESPI Report 53.

[30] GNS, Galileo Open Service, Service Definition Document (2019) 1.6. https://galileognss.eu/wp-content/uploads/2020/08/Galileo-OS-SDD_v1.1.pdf.

controlled access and authentication function to the users".[31] The CAS will provide greater accuracy than OS and will be a fee-based service.

3 Cooperation in the Field of Satellite Navigation

In terms of radiofrequency compatibility, the International Telecommunication Union (ITU) framework is essential. ITU, a specialized agency of the United Nations, coordinates the use of radio spectrums, which are limited resources, anddevelops recommendations to aid compatibility. Additional Radio Navigation Satellite Service (RNSS) spectrums were allocated during the ITU World Radiocommunication Conference (WRC 2000) and long term sustainable development of new RNSS and protection of RNSS existing systems were guaranteed during the WRC in 2003.[32] As underlined in the Article 4(10) of Radio Regulations,[33] "Member States recognize that the safety aspects of radionavigation and other safety services require special measures to ensure their freedom from harmful interference; it is necessary therefore to take this factor into account in the assignment and use of frequencies".

In terms of interoperability among global and regional systems, the International Committee on GNSS (ICG) and the GNSS Providers Forum are centrally important. The first of these, established on a voluntary basis in 2005 under the auspices of the United Nations, is an international forum for discussion of navigation issues. It promotes voluntary cooperation on matters of mutual interest related to civil satellite-based navigation services. Consequently, the principle of functioning of the Committee is based on a 'best practices' approach.[34] As of December 2019, members of the ICG are Australia, China, India, Italy, Japan, Malaysia, New Zealand, Nigeria, the Russian Federation, the United Arab Emirates, the United States of America and the EU (and other intergovernmental and non-governmental organizations).[35] The second, including GNSS and RNSS providing States, established in 2007 under the umbrella of the United Nations, aims to promote greater compatibility and interoperability among current and future providers. The GNSS Providers Forum "also acts as a mechanism to continue discussions on important issues addressed by ICG that require inputs from system providers".[36] The United Nations Office for Outer Space Affairs is the Executive Secretariat of the ICG and the GNSS Providers Forum.

---

[31] Ibid.

[32] See generally Yvon Henri, Attila Matas, "RNSS and the ITU Radio Regulations" (January/February 2018) Inside GNSS 34–35.

[33] ITU, Radio Regulations (no 7).

[34] Aliberti, Lahcen (no 11) 31.

[35] Committee on the Peaceful Uses of Outer Space, Fourteenth meeting of the International Committee on Global Navigation Satellite Systems, Distr.: General 20 December 2019, A/AC.105/1217, Annex I.

[36] Ibid., para.3.

The use of GNSS is essential in order to ensure the safety in the air and in the sea. Therefore, the International Civil Aviation Organization (ICAO) and the International Maritime Organization (IMO)[37] have been focusing on the use of GNSS applications in aviation and maritime navigation. These organizations are formulating the principles and standards of GNSS use in aviation and maritime operations.

Charter on the rights and obligations of states relating to GNSS services issued by the ICAO Assembly in 1998 provides the principles applicable in the implementation and operation of GNSS.[38] International Standards and Recommended Practices (SARP) and procedures for air navigation services have been developed by the GNSS panel (GNSSP) and were published in November 2001 as part of Annex 10 to the Convention on Civil Aviation.[39] In addition to these standards and practices, the GNSS Manual, prepared by the GNSSP, aims "to provide information about GNSS technology and operational applications to assist State regulators and Air Navigation Service (ANS) providers to complete the safety and business case analyses needed to support implementation decisions and planning".[40]

In 1997, the IMO issued a maritime policy for the future GNSS, giving maritime requirements for a GNSS (which could be reassessed and updated on the basis of new developments).[41] Through the revised version of Chapter V of The International Convention for the safety of life at sea in 2000 (entered into force in 2002) carrying a GNSS or terrestrial radionavigation receiver is required on all ships throughout their voyage. In 2015, the IMO adopted the Performance Standards for multi-system (using signals from two or more GNSS), shipborne radionavigation receivers allowing for improved position, velocity and time data.[42] They also recommended that governments to ensure that these receivers installed on, or after 31 December 2017, conform to performance standards.

---

[37] See generally Hiroyuki Yamada, "IMO and the GNSS: Navigating the Seas" (September/October 2017) Inside GNSS 40–44.

[38] ICAO, Charter on the rights and obligations of states relating to GNSS services, A32-19 adopted at the 32nd session of the Assembly, 2 October 1998, www.icao.int/Meetings/AMC/MA/Assembly%2032nd%20Session/resolutions.pdf.

[39] ICAO, International Standards and Recommended Practices and procedures for Air navigation services, Annex 10 to the Convention on Civil Aviation, Sixth Edition, October 2001. www.icao.int/Meetings/anconf12/Document%20Archive/AN10_V2_cons%5B1%5D.pdf.

[40] ICAO, Global Navigation Satellite System (GNSS) Manual, Second Edition—June 2012, Executive summary para 5 www.icao.int/Meetings/anconf12/Documents/Doc.%209849.pdf. Third edition of this Manual was published in 2017.

[41] IMO, Maritime Policy for a Future GNSS, Res. A860(2), 27 November 1997 https://wwwcdn.imo.org/localresources/en/KnowledgeCentre/IndexofIMOResolutions/AssemblyDocuments/A.860(20).pdf.

[42] IMO, Performance Standards for Multi-System Shipborne Radionavigation Receivers, Res. MSC.401(95), 8 June 2015) https://wwwcdn.imo.org/localresources/en/KnowledgeCentre/IndexofIMOResolutions/MSCResolutions/MSC.401(95).pdf.

4 International Cooperation Development and Implementation

On an international dimension, the European Union pursues three approaches to cooperation: multilateral, bilateral and regional.

4.1 Multilateral

On the multilateral level, the European Union is a member of the ICG. It is an active participant of the GNSS Providers Forum. It is also involved in the activities of World Radio-communication Conferences and of the meetings of the ITU study groups and working groups.[43] Galileo has been used in aviation and maritime navigation fields, which has led the European Union to participate in the activities of the IMO and ICAO. In 2016, Galileo was recognized by the Maritime Security Committee of the IMO as a component part of the World Wide Radio Navigation System (WWRNS).[44] Galileo SAR service is compliant with ICAO SAR requirements, effective on 11 July 2016 and applicable from 1 January 2021, in addressing the Global Aeronautical Distress & Safety System and Autonomous Distress Tracking.[45] New SARPs adopted by the ICAO in November 2020 have created a basis for the development of new Dual Frequency Multi-Constellation Satellite Based Augmentation Systems enabling signal corrections from multiple GNSS (so not only from GPS and GLONASS as is currently the case).[46] The Union actively contributes to committees, commissions and working groups of the organizations cited above.

---

[43] UNOOSA, "International Committee on Global Navigation Satellite Systems, The way forward, 10 years of achievement 2005–2015" (no 8) 33–34.

[44] Maritime Safety Committee (MSC), 96th session, 11–20 May 2016, https://imo.org/en/MediaCentre/MeetingSummaries/Pages/MSC-96th-session.aspx. GPS and GLONASS were recognized as part of the WWRNS in 1996, BeiDou in 2014. See generally Yamada (no 37) 42–43.

[45] GNS, "Galileo Supports Compliance with ICAO SAR requirements", 4 November 2018, www.gsa.europa.eu/newsroom/news/galileo-supports-compliance-icao-sar-requirements.

[46] GNS, "International Civil Aviation Organization Navigation System Panel approves new Standards and recommended practices putting forth the use of EGNOS and Galileo", 2 December 2020, www.gsa.europa.eu/newsroom/news/international-civil-aviation-organisation-approves-new-standards.

4.2 Bilateral

4.2.1 With GNSS/RNSS Providing States

The United States of America

Even though the US was initially opposed to the Galileo Programme, especially for security reasons, it finally withdrew its opposition at the end of discussions, during which a compromise was found, alleviating the concerns of both parties. The US, the EU, and its Member States signed the Agreement on the Promotion, Provision and Use of Galileo and GPS Satellite-based Navigation Systems and Related Applications in 2004.[47] One of the main aims of this agreement is to guarantee interoperability and compatibility of these systems.[48] Four working groups have been established by this agreement: a working group on radio frequency compatibility and interoperability; a working group on trade and civil applications; a working group to promote cooperation on the design and development of the next generation of civil satellite-based navigation and timing systems and a working group on security issues.[49] This agreement was supplemented by a Joint Statement dated 23 October 2008 on the use of a common GPS-Galileo signal, ensuring interoperability.[50] Galileo is currently compatible and fully interoperable with GPS.[51]

With the waivers concerning Galileo E1 and E5 signals approved by the Federal Communications Commission (FCC) on 15 November 2018, the users in the US have been permitted to use these signals for non-federal purposes.[52] Consequently, these signals (E1 and E5) can be used in combination with GPS. The decision of the FCC would improve the "availability, reliability, and resiliency" of navigation service in the US.[53] Since then, important players in the american market such as Broadcom and Qualcomm Technologies have offered solutions integrating Galileo, and a significant increase in usage has been seen in Galileo users in the US.[54]

---

[47] Agreement on the promotion, provision and use of Galileo and GPS satellite-based navigation systems and related applications between the European Community and its Member States, of the one part, and the United States of America, of the other part, signed at Dromoland Castle, Co. Clare, on the twenty-sixth day of June 2004.

[48] Ibid., Article 4.

[49] Ibid., Article 13.

[50] Joint Statement on GPS and Galileo Cooperation by representatives of the United States of America, the European Community and its Member States, published 23 October 2008, www.gps.gov/policy/cooperation/europe/2008/joint-statement/.

[51] GNS, FAQ, www.gsc-europa.eu/galileo/faq#operate.

[52] Dee Ann Davis, "FCC poised to Approve broad use of Galileo in the US", (InsideGNSS, 15 November 2018) https://insidegnss.com/fcc-poised-to-approve-broad-use-of-galileo-in-u-s/.

[53] Ibid.

[54] InsideGNSS, "GNSS Chip manufacturers expect big Galileo roll out in US" (18 March 2019) https://insidegnss.com/gnss-chip-manufacturers-expect-big-galileo-roll-out-in-u-s/.

China

In 2003, an agreement between China and the EU was signed to encourage, facilitate and enhance cooperation.[55] In this agreement, China associated with Galileo, and became a member of the Galileo Joint Undertaking (GJU). After the liquidation of the GJU and the exclusion of China from the key decision-making process, China decided to build its own global system.[56] This separation and the conflict over the PRS frequency overlay issue negatively influenced EU-China relationships for a long time.[57] The conflict stemmed from China's decision to use the same spectrum used by PRS for its highly encrypted service. However, this overlap with PRS "will not impinge on the operations of either system but will make it difficult for either one to jam the signals of the other in the event of a conflict".[58] A Joint Statement on Space Technology Cooperation was signed between the parties as an annex to a broader EU-China summit held on 20 September 2012 in Brussels.[59] In this Statement, each party expressed their willingness for a resolution to the Galileo/EGNOS and Compass frequency issue under the ITU Framework. A working group on compatibility and interoperability between the BeiDou and Galileo has been established "to continuously promote frequency coordination".[60] A joint statement on the compatibility and interoperability of BeiDou/GPS and of BeiDou /GLONASS has been signed, but none has yet been signed for BeiDou/Galileo.[61]

The Russian Federation

In 2006, the EU and Russia initiated a dialogue in order to resolve the issues of compatibility and interoperability of their global systems.[62] Since the annexation of Crimea by Russia in 2014, the EU's relations with Russia relations have been tense, and cooperation in the field of satellite navigation has been at a low-level.[63]

---

[55] Cooperation agreement on a civil global navigation satellite system (GNSS) between the European community and its member states and the People's Republic of China, Beijing, 30 October 2003.

[56] Dejian Kong, Fabio Tronchetti, "State of play in China-strategy, governance, policy and law of the Beidou navigation satellite system" (January/February 2017) Inside GNSS 38–39.

[57] Spyros Pagkratis, "Space policies, issues and trends in 2009/2010" (2010) 23 ESPI Report 100.

[58] Peter de Selding, "China and Europe taking their navigation dispute to ITU", (Space News, 8 October 2012) https://spacenews.com/china-and-europe-taking-their-navigation-dispute-itu/.

[59] Ibid.

[60] China Satellite Navigation Office, Development of the BeiDou Navigation Staellite System (Version 3.0), (December 2018) 20. www.beidou.gov.cn/xt/gfxz/201812/P020181227529626058961.pdf.

[61] Cao Qingqing, "Expert: BeiDou eyes win–win cooperation with other global navigation systems" (CGTN, 20 December 2019) https://news.cgtn.com/news/2019-12-20/Beidou-eyes-win-win-cooperation-with-other-global-navigation-systems-MzTWCZywnK/index.html.

[62] Sitruk, Plattard (no 3) 53.

[63] Peter Gutierrez, "EU and Russia: lost in space" (November/December 2014) Inside GNSS 27–31.

Discussions between the EU and Russia, for a future cooperation agreement related to satellite navigation, are still ongoing.[64]

Japan

Frequency compatibility between Galileo and QZSS has been achieved under the framework of the ITU.[65] A Cooperation Arrangement was signed between the Government of Japan and the European Commission on 8 March 2017 "to enhance EU-Japan policy cooperation in order to prioritise industrial sectors for utilising satellite positioning and creating new business services".[66] As a result of this, Japan and Japanese industry have a very close relationship with Europe (and European industry) regarding downstream services in the field of GNSS.[67]

India

In 2005, an agreement between India and Europe had been signed for the participation of India in the Galileo programme.[68] India, unhappy about not being considered to be a full partner in this agreement, decided to withdraw from Galileo and to acquire its own RNSS.[69] Frequency compatibility between Galileo and NavIC has been achieved under the framework of the ITU.[70] India is an important partner of Europe in downstream GNSS industrial cooperation especially in the context of GNSS.asia, an international effort funded by the EU.[71]

---

[64] UNOOSA, "International Committee on Global Navigation Satellite Systems, The way forward, 10 years of achievement 2005–2015" (no 8) 17.

[65] GPS World, "Directions 2016: Galileo – strategic tool for European autonomy' (28 December 2015) www.gpsworld.com/directions-2016-galileo-strategic-tool-for-european-autonomy/.

[66] GSA, "Japan joins GNSS table with QZSS", Published 6 September 2017 www.gsa.europa.eu/newsroom/news/japan-joins-gnss-table-qzss.

[67] ESPI Report 74 - Securing Japan - Full Report (2020) 100.

[68] European Commission, "The Galileo family is further expanding: EU and India seal their agreement", 7 September 2005, https://ec.europa.eu/commission/presscorner/detail/en/IP_05_1105.

[69] ESPI Report 69, "Europe-India Space Cooperation: Policy, Legal and Business Perspectives from India" (2019) 66.

[70] GPS World (no 65).

[71] ESPI Report 69 (no 69) 69–70.

4.2.2 With Non-GNSS/RNSS Providing States

General Points

The EU has also had close relations with non-GNSS providing States. The inclusion of third parties in the programme "shar[ing] the European Union's interests in promoting it nationally and internationally, has resulted in a reduction in the technical and political risks involved".[72] Cooperation agreements were signed with the following States: Israel,[73] Ukraine,[74] South Korea,[75] and Morocco.[76] The main goal of these agreements is to "encourage, facilitate and enhance cooperation between the parties in civil global satellite navigation". Scope of cooperative activities include generally scientific research, industrial manufacturing, training, application, service and market development, trade, radio-spectrum issues, integrity issues, standardization and certification and security (Article 4 in each of the Agreements). Implementation of a ground regional augmentation system based on Galileo in these States is also one of the features of the cooperative activities (Article 11 in each of the Agreements).

Norway[77] and Switzerland,[78] which are non-EU states but full members of ESA, also joined Galileo. Norway accepted the deployment, maintenance and replacement of two ground facilities in the territories under its jurisdiction (Article 5). Norway has demonstrated its interest in accessing PRS signals (Article 9). Switzerland too has expressed its interest in accessing PRS signals subject to a separate agreement (Article 15). Switzerland's participation in GSA (after the conclusion of an agreement between the European Union and Switzerland) is provided for by Article 16. Switzerland's representatives shall also participate, as observers, in activities of certain Committees such as the GNSS Programme Committee and the GNSS Security Board (Article 17). It should be stipulated that the atomic clocks used in the Galileo satellites have been provided by a Swiss company.

---

[72] Ailio (no 6) 206.

[73] Cooperation agreement on a civil global navigation satellite system (GNSS) between the European community and its member states and the state of Israel, Brussels, 13 July 2004.

[74] Cooperation agreement on a civil global navigation satellite system (GNSS) between the European Community and Ukraine, Kiev, 1 December 2005.

[75] Cooperation agreement on a civil global navigation satellite system (GNSS) between the European community and its member states and the republic of Korea, Brussels, 19 October 2006.

[76] Cooperation agreement on a civil global navigation satellite system (GNSS) between the European community and its member states and the kingdom of Morocco, Brussels, 12 December 2006.

[77] Cooperation agreement on satellite navigation between the European Union and its member states and the Kingdom of Norway, Brussels, 22 September 2010.

[78] Cooperation agreement between the European Union and its member states and the Swiss Confederation on the European satellite navigation programmes, Brussels, 18 December 2013.

The Special Case of the United Kingdom

After Brexit, the UK remains an ESA member state. However, the UK's participation in EU flagship programmes such as Galileo and Copernicus may be adversely affected.[79] The UK played a central role in design, development and management of Galileo and has contributed 12% ($1.5 billion) to the total budget. After Brexit, the UK won't have access to PRS in the absence of a security agreement with the EU, and British companies won't be able to participate in the R&D projects of Galileo.

Appreciating that developping a sovereign UK GNSS system is extremely difficult to realize because of the high cost (notwithstanding the capabilities of the local industry), the UK government is seeking 'newer, more innovative' approaches through Space-Based Positioning Navigation and Timing Programme.[80] The UK's current use of the encrypted signal of GPS is not affected by Brexit. The UK could have access to PRS signals if it were to complete and sign a separate agreement with the EU, but the UK has not yet requested access to this service.[81]

Security related aspects constitute an important chapter of the new UK-EU partnership. According to the negotiations directives for a new partnership proposed by the EU Commission and approved by the EU Council in February 2020,[82] the envisaged partnership should provide for the possibility for the United Kingdom to access to the PRS (para. 135). For example, the UK may use PRS signals "for sensitive applications in the context of Union operations or ad hoc operations involving its Member States". The UK's access to PRS should be conditional upon the United Kingdom "participating in the non-security related activities of the Union's Space Programme (…) unless and until the United Kingdom grants the Union access to the envisaged United Kingdom Global Navigation Satellite System" (para.136(b)). Accordingly, the use of the PRS signals by the UK should not prejudice the essential security interests of the Union and its Member States (para.136 (a)).

---

[79] On the effects of Brexit for the British space industry see generally Joanna Wheeler, "The Consequences Post Referendum for the UK Satellite and Space Industry" (2016) 41 Air & Space Law 445–458.

[80] Jeff Foust, "U.K. to revise strategy for satellite navigation system" (Space News, 25 September 2020) https://spacenews.com/u-k-to-revise-strategy-for-satellite-navigation-system/.

[81] Peter Gutierrez, "The rocky road to a UK GNSS" (Inside GNSS, 30 November 2020) https://insidegnss.com/the-rocky-road-to-a-uk-gnss/.

[82] Council decision, authorising the opening of negotiations for a new partnership with the United Kingdom of Great Britain and Northern Ireland, Brussels, 3.2.2020 https://eur-lex.europa.eu/legal-content/EN/TXT/?uri=CELEX%3A52020PC0035.

4.3 Regional

The last approach pursued by the Union is the regional one, which is based on "awareness raising, training and demonstration activities in order to inform on the opportunities and benefits of Galileo".[83] South East Asia, Latin America, and Africa/the European neighborhood are target regions. Among these, the Asian one is the most important and a hotspot for Galileo.[84] This has led Europe to focus its efforts on the Asian market. The main objectives of the EU Projects are to promote the use of Galileo in these regions by raising awareness and capacity building (1), to prepare the market for the penetration of Galileo applications and services in these regions (2), and to enhance and implement industrial cooperation between the Union and Asian countries in the field of GNSS (3). The main projects engaged by the GSA introducing Galileo in the following regions are:

South East Asia

- **GNSS.asia Project** (Cross-continental Cooperation with GNSS.asia)[85];
- **BELS Project** (Building European Links toward South East Asia in the field of GNSS)[86];
- **SEAGAL Project** (South East Asia centre on European GNSS for international cooperation and Local Development)[87];
- **G-Navis Project** (Growing Navis Centre located in Hanoi, Vietnam).[88]

Latin America

- **GACELA Project** (Galileo Centre of excellence for Latin America)[89];
- **ENCORE Project** (Enhanced code Galileo receiver for land management in Brazil)[90];

---

[83] Ailio (no 6) 207.

[84] GSA, "GNSS Market Report" (no 20) 6,12.

[85] GSA, "Cross-continental cooperation with GNSS.asia" (Published 19 February 2015) www.gsa.europa.eu/news/cross-continental-cooperation-gnssasia.

[86] GSA, "Building European Links toward South East Asia in the field of GNSS" www.gsa.europa.eu/building-european-links-toward-south-east-asia-field-gnss.

[87] GSA, "South East Asia Centre on European GNSS for international cooperation and local development" www.gsa.europa.eu/south-east-asia-centre-european-gnss-international-cooperation-and-local-development-0.

[88] GSA, "Growing Navis", www.gsa.europa.eu/growing-navis.

[89] GSA, "Galileo Centre of excellence for Latin America", www.gsa.europa.eu/galileo-centre-excellence-latin-america-0.

[90] GSA, "Enhanced code Galileo receiver for land management in Brazil", www.gsa.europa.eu/enhanced-code-galileo-receiver-land-management-brazil.

- **CALIBRA Project** (Countering GNSS high Accuracy applications Limitations due to Ionospheric disturbances in Brazil).[91]

Africa/European Neighborhood

- **MAGNIFIC Project** (Multiplying in Africa European Global Navigation Initiatives Fostering Interlaced Cooperation)[92];
- **AiA Project** (Awareness in Africa: disseminating knowledge on EGNOS and Galileo in Africa to foster local development)[93];
- **SATSA Project** (SBAS Awareness and Training for South Africa)[94];
- **GEMCO Project** (Galileo EuroMed Cooperation Office located in Tunis)[95].

5 Concluding Remarks

In the field of satellite navigation, States are between competition and cooperation.[96] A priori these two concepts seem contradictory. It is obvious that the space domain is a sensitive domain from the point of view of the national security of States. In an increasingly competitive context, cooperation is seen as a constraint. I believe that these two concepts are not always contradictory, and that cooperation can also serve national/regional interests. As put forward by the ESA's Director Johann-Dietrich Wörner, "competition is a driver, cooperation is an enabler".

Owing to "inappropriate choices with respect to a governance scheme and financing mechanisms",[97] Europe has arrived late in "the GNSS Universe". Good management of Galileo and optimizing the benefits of the programme require a close relationship between the EU, Member States and the ESA.[98] The temporary interruption of service for several days in July 2019, highlighted once again problems related

---

[91]GSA, "Countering GNSS high Accuracy applications Limitations due to Ionospheric disturbances in Brazil", www.gsa.europa.eu/countering-gnss-high-accuracy-applications-limitations-due-ionospheric-disturbances-brazil.

[92]GSA, "Multiplying in Africa European global navigation initiatives fostering interlaced cooperation", www.gsa.europa.eu/multiplying-africa-european-global-navigation-initiatives-fostering-interlaced-cooperation.

[93]GSA, "Awareness in Africa: disseminating knowledge on EGNOS and Galileo in Africa to foster local development", www.gsa.europa.eu/awareness-africa-disseminating-knowledge-egnos-and-galileo-africa-foster-local-development.

[94]GSA, "SBAS awareness and training for South Africa" www.gsa.europa.eu/sbas-awareness-and-training-south-africa.

[95]GSA, "GEMCO takes off in Tunis", www.gsa.europa.eu/news/gemco-takes-tunis. Euromed countries include Algeria, Egypt, Israel, Jordan, Lebanon, Libya, Morocco, Palestine, Syria, Tunisia and Turkey.

[96]Volynskaya (no 1) 376.

[97]Giannopapa (no 16) 7.

[98]Council of the European Union, "Space as an enabler" (no 14) paras. 15,16,19.

to the governance structure of Galileo.[99] Europe should find efficient cooperative solutions to develop Galileo governance.

Through its diplomacy, Europe should promote the use of Galileo and create new opportunities for European industry in emerging markets. Indeed, the use of diplomacy is an important tool to ensure the competitiveness of European space industry.[100] Industrial cooperation is vital in enabling market access for European industry. The EU should continue to ensure that chipset and receiver manufacturers build Galileo enabled devices. The EU should also develop more cooperative ventures in the field of satellite navigation especially with South Asian countries.[101]

With the emergence of New Space, the space market is increasingly competitive. European R&D projects (such as Horizon 2020 and Horizon Europe) are vital tools to support the competitiveness of the European GNSS industry. Europe should continue to support academic and research activities. Entrepreneurship should also be encouraged, and a more favorable environment should be provided to entrepreneurs at the legal and financial levels.

The EU should contribute more to build the capacity of developing countries in using GNSS and to raise awareness of Galileo in these countries through the organization of workshops. The Union should build up sustainable cooperation with these countries for which social and economic benefits of GNSS technology are crucial. These countries should be assisted in order to integrate GNSS into their national infrastructure.

Realization of navigation services at the user level, requires an international cooperation of GNSS and RNSS providers in order to ensure compatibility and interoperability. Ensuring compatibility and interoperability with other systems is the main goal of European Space objectives. Europe should continue to advance cooperation and system interoperability between Galileo and other systems.

Tugrul Cakir is a lecturer specialised in Space Law at the Ankara Yildirim Beyazit University since March 2020. He earned a master's degree (2014) and a Ph.D. degree (2019) at the Université Jean Moulin Lyon III. His master's thesis and Ph.D. thesis treated different aspects of International and National Space Law. His research outputs, among others, include a chapter relating to sources of Space Law published in December 2020 by the Hague Academy of International Law in a collected work (*50 Years of Space Law-Space Law in 50 Years*).

---

[99] Inside GNSS, "Lessons to be learned from Galileo signal Outage" (Inside GNSS, 1 October 2019) https://insidegnss.com/lessons-to-be-learned-from-galileo-signal-outage/.

[100] ESPI, "European Space Strategy in a Global Context - Executive Summary" (2020) 75 ESPI Report 9.

[101] See generally Jana Robinson, Fabien Evrard, "Status of Europe's Space Cooperation with Asia" (2012) 61 ESPI Perspectives.

Rise of Mega Constellations: A Case to Adapt Space Law Through the Law of the Sea

Lauryn Hallet

Abstract A phenomenon is on the rise, one of mega constellations. Around 190 constellations are under progress, and SpaceX alone plans launch over 40,000 small satellites as part of Starlink. By January 2021, Starlink had added 900 systems to the 6,250 still in space. Reports on collision risks and uncontrolled reentries evidence the need to settle the debate on the lower boundary of outer space. Also, mega constellations and associated events such as the escalation of space activities and launches, the proliferation of private actors and the intensification of global competition, will inevitably exacerbate the flaws of current registration and enforcement mechanisms provided by space law. If law does not meet the needs and realities of new trends and markets stemming from mega constellations with adapted legal frameworks, unfortunate consequences shall be expected. It is demonstrated that the best way out of stalemate is to take the law of the sea as an example through analogies between outer space and the sea.

1 Introduction

Around 190 constellations are under development; amongst those, several mega constellations. For instance, SpaceX has received the green light to launch 12,000 satellites in low Earth orbit (LEO) and plans to add 30,000 more.[1] Considering there are already 6,250 satellites orbiting Earth, millions of debris objects, most in LEO, this is a problem. The repercussions on safety and the environment could be disastrous to a point of no-return if the Kessler effect was to be triggered. In this

---

[1] Caleb Henry, 'SpaceX submits paperwork for 30,000 more Starlink satellites' (SpaceNews, 15 October 2019) https://spacenews.com/spacex-submits-paperwork-for-30000-more-starlink-satellites/ (all websites cited in this publication were last accessed and verified on 19 January 2021).

This article is an adapted version of a thesis submitted to the University of Bristol in the frame of a Master of Law degree in International Law for the Faculty of Social Sciences and Law.

L. Hallet (✉)
University of Bristol, Bristol, UK

context, questions of heightening collision risks and reentries should evidence the need to settle once and for all the debate on the lower boundary of outer space. Mega constellations and associated phenomena such as the escalation of space activities and launches, the proliferation of private actors and the intensification of competition, will inevitably exacerbate the flaws of current registration and liability mechanisms provided through space law.

This article investigates how space law should be adapted to take into account new trends such as mega constellations, and how it can do so through the law of the sea. It is based on the premise that in view of the problematic stagnation of international space law, the law of the sea could offer useful guidance. This assertion is based on the fact that both legal areas regulate environments exhibiting significant commonalities. The rationale is therefore to make analogies between similar problematics and the way these have been treated. Often, the analogy between space law and the law of the sea, or between outer space and the high seas/deep seabed has raised debate and even some eyebrows.[2] As Mačák has stated, space activities are of such a nature that they require absolutely novel solutions and any analogies to the sea should be rejected.[3] As a response to this statement, it should be emphasized that the aim of analogies, at least for the purposes of this article, is not to take the law of the sea at face value and copy and paste it onto space law, but to make an empirical analysis of similar circumstances, because such an analysis is not yet possible for many space problematics, and the object of the exercise is to anticipate and prevent challenges.[4]

In the past, the law of the sea has revealed to be relevant and has had a non-negligible influence on the development of space law and its principles.[5] This is most notably the case for the freedom of the sea and the freedom of space exploration, and the concept of peaceful purposes.[6] At some point, the law of the sea and space

---

[2] Elizabeth Mendenhall, 'Treating Outer Space Like a Place: A Case for Rejecting Other Domain Analogies' (2018) 16(2) *Astropolitics* 97; Nina Tannenwald, 'Law versus Power on the High Frontier: The Case for a Rule-Based Regime for Outer Space' (2004) 29 *Yale Journal of International Law* 363, 364; Kubo Mačák, 'Silent War: Applicability of the *Jus in Bello* to Military Space Operations' (2018) 94 *International Law Studies* 1, 17–18.

[3] Kubo Mačák, 'Silent War: Applicability of the *Jus in Bello* to Military Space Operations' (2018) 94 *International Law Studies* 1, 17–18.

[4] Armel Kerrest, 'Space law and the law of the sea', in Christian Brünner and Alexander Soucek, *Outer Space in Society, Politics and Law* (2011, Springer Wien New York), 247.

[5] J. E. S. Fawcett, *Outer Space: New Challenges to Law and Policy* (OUP 1984), 7; Armel Kerrest, 'Space law and the law of the sea', in Christian Brünner and Alexander Soucek, *Outer Space in Society, Politics and Law* (2011, Springer Wien New York), 247; Frans G. von der Dunk, 'A Tale of Two Oceans: Governance of Terrestrial and Outer Space "Global Commons"' (2012) 2(1) *Asian Journal of Air and Space Law* 31, 32; William V. Dunlap, 'International Boundaries: The Next Generation' (1999–2000) *IBRU Boundary and Security Bulletin* 106, 107, 109.

[6] United Nations Convention on the Law of the Sea (UNCLOS) (opened for signature 10 December 1982, entered into force 16 November 1994) 1833 UNTS 3, arts 87, 88, 141, 143; Treaty on Principles Governing the Activities of States in the Exploration and Use of Outer Space, including the Moon and Other Celestial Bodies (Outer Space Treaty) (opened for signature 27 January 1967, entered into force 10 October 1967) 610 UNTS 8843, arts I, IV; Agreement governing the Activities of States on the Moon and Other Celestial Bodies (Moon Agreement) (opened for signature 18 December 1979, entered into force 11 July 1984) 1363 UNTS 23002, arts 3, 6; Yoshifumi Tanaka, *The International*

law even concurrently developed regimes for areas beyond national jurisdiction—the Area and the Moon and other celestial bodies—to the extent that a parallel vision can be observed. In particular, the infamous principle of "common heritage of mankind" can be found in both the Moon Agreement and the United Nations Convention on the Law of the Sea (UNCLOS).[7] Outer space and the sea share many more common attributes and, as such, face similar issues. The law of the sea has reached a certain level of complexity and sophistication over time, and through trial and error, has already addressed and solved many issues, such as the settlement of maritime zones. Unfortunately, sometimes it has failed to foresee potential risks and still struggles to deal with these, for example with regard to flags of convenience.[8] However, space law could learn as much from the triumphs and strengths of the law of the sea as from its failures and flaws.

Ultimately, making analogies between the sea and space in order to help the law adapt to new situations is more important than some may be inclined to admit.[9] As has rightly been said by International Tribunal for the Law of the Sea (ITLOS) President Jesus, the law should cater to the needs of society, but to do so, it has to adapt to evolutions and trends in the field it regulates, and be careful to prevent or contain harmful tendencies that may arise.[10]

Space law is relatively new, especially in comparison with the law of the sea where practices and laws with regard to various issues are much older and have gone through many stages, developments and innovations, which is why space law could benefit from the examples of the law of the sea.[11]

Development and key elements of The International Law of the Sea (2.1) and International Space Law (2.2) are summarized and reviewed, before discussing the analogies between space and the sea. Two themes have been chosen because they

---

Law of the Sea (2nd edn, CUP 2015), 16; William V. Dunlap, 'International Boundaries: The Next Generation' (1999–2000) *IBRU Boundary and Security Bulletin* 106, 107, 107, 110; H. A. Wassenbergh, 'Parallels and Differences in the Development of Air, Sea and Space Law in the Light of Grotius' (1984) 9 *Annals of Air Space Law* 163, 172.

[7]UNCLOS, arts 125, 136, 150, 155, 311; Agreement governing the Activities of States on the Moon and Other Celestial Bodies (Moon Agreement) (opened for signature 18 December 1979, entered into force 11 July 1984) 1363 UNTS 23002, art 11; Nina Tannenwald, 'Law versus Power on the High Frontier: The Case for a Rule-Based Regime for Outer Space' (2004) 29 *Yale Journal of International Law* 363, 375; William V. Dunlap, 'International Boundaries: The Next Generation' (1999–2000) *IBRU Boundary and Security Bulletin* 106, 110.

[8]Armel Kerrest, 'Space law and the law of the sea', in Christian Brünner and Alexander Soucek, *Outer Space in Society, Politics and Law* (2011, Springer Wien New York), 249.

[9]Nina Tannenwald, 'Law versus Power on the High Frontier: The Case for a Rule-Based Regime for Outer Space' (2004), 29 *Yale Journal of International Law* 363; Armel Kerrest, 'Space law and the law of the sea', in Christian Brünner and Alexander Soucek, *Outer Space in Society, Politics and Law* (2011, Springer Wien New York), 248.

[10]José Luis Jesus, 'Protection of Foreign Ships against Piracy and Terrorism at Sea: Legal Aspects' (2003) 18 *The International Journal of Marine and Coastal Law* 363, 381–382, in Natalie Klein, *Maritime Security and the Law of the Sea* (OUP 2013), 11–12.

[11]Armel Kerrest, 'Space law and the law of the sea', in Christian Brünner and Alexander Soucek, *Outer Space in Society, Politics and Law* (Springer, 2011), 247.

exemplify the imperativeness of prompt resolutions and the potential consequences of prolonged inaction. The corresponding sections are structured according the same *modus operandi*. First, the situation with regard to outer space and a certain space law problematic is described. Then, the same situation in the context of the sea is discussed. How the law of the sea has tackled the issue is examined and potential flaws or strengths are highlighted. This is followed by a reflection on whether the strengths may serve as a stepping stone for space law to develop its own regime in regard to the problem at issue, or whether the flaws could serve as warning signs that will enable space law to anticipate future downfalls. In other words, the goal is to transpose past lessons learned in the context of the law of the sea to the context of space law, not to transpose the law itself.

The section on boundary delimitation draws parallels between the airspace-outer space boundary and the delimitation of maritime zones, and focuses on the boundary between the territorial sea and the high seas. It is demonstrated that if the outer limit of airspace and its height can follow the same logic as the outer limit of the territorial sea and its breadth, the manner in which the baseline of the latter is established is not suitable for outer space.

The next section evokes the probable eruption of flags of convenience in outer space caused by an improper registration and liability system that allows for potentially great disparities amongst individual states. This is linked with the corresponding issue of flags of convenience at sea. It is concluded that the law of the sea has failed to act appropriately against the proliferation of flags of convenience. As such, space law should prevent the registration and liability regimes to facilitate an outbreak of "registries of convenience" in outer space. It could do so by reproducing measures initiated in the law of the sea, while improving them to fill the vacuum left in the latter and do so before the actual eruption of flags of convenience.

In conclusion, this article argues that the law of the sea does offer guidelines for space law in its development and adaption to the circumstances of the modern world.

2 Developments

2.1 *The International Law of the Sea*

The international law of the sea regulates activities conducted at sea.[12] The sea presents many strategic opportunities which are beyond the realm of commerce.[13] Indeed, dominion over the oceans has granted considerable military advantages.[14] Consequently, in order to obtain and retain the commercial and military high ground,

---

[12] James Harrison, *Making the Law of the Sea: A Study in the Development of International Law* (CUP 2011), 1.

[13] Natalie Klein, *Maritime Security and the Law of the Sea* (OUP 2013), 13; Donald Rothwell and Tim Stephens, *The International Law of the Sea* (2nd edn, Hart Edition 2016), 2.

[14] Ibid.

it was in the interest of states to support freedom of navigation without any obstruction so as to enable the easy export of goods and transport of people on the one hand, and the free movement of naval fleets in key locations on the other hand.[15]

This in turn gave rise to *mare liberum*, a concept described by Grotius in 1609, which portrays the sea as a limitless commons, free for use by all, barred from possession by any one.[16] However, by 1635, the opposite principle of *mare clausum*, developed by John Selden, had emerged.[17] The aim of Selden's work was to endorse and demonstrate the existence of a state practice where individual states would assert their sovereignty over parts of the oceans.[18] While absolute freedom was not unanimous, it did eclipse *mare clausum* for a moment, and its contemporary application is found in Article 87 of the UNCLOS.[19] However, the debate was not indefinitely closed. Indeed, it was raised anew by security concerns and the potentiality for naval powers to control parts of the sea, or approach and use force against coastal states while still complying with the principle.[20]

Accordingly, in contrast with the concept of the high seas and those new concerns, the concept of "territorial sea" emerged, in terms of which coastal states generally asserted security and economic rights.[21] Eventually, this became common practice. But, despite the proliferation of claims to territorial seas, no general agreement could be reached as to its breadth.[22] This is in part the reason for the failure to codify the

---

[15] Natalie Klein, *Maritime Security and the Law of the Sea* (OUP 2013), 13; Yoshifumi Tanaka, *The International Law of the Sea* (2nd edn, CUP 2015), 17; Malcolm Evans, 'Law of the Sea', in Malcolm Evans (ed), *International Law* (4th edn, OUP 2015), 651–652; Malcolm Shaw, *International Law* (8th edn CUP 2017), 410; Douglas Guilfoyle, 'The High Seas', in Donald Rothwell et al. (eds), *The Oxford Handbook of the Law of the Sea* (OUP 2015), 203.

[16] Natalie Klein, *Maritime Security and the Law of the Sea* (OUP 2013), 12–13; Yoshifumi Tanaka, *The International Law of the Sea* (2nd edn, CUP 2015), 17; Donald Rothwell and Tim Stephens, *The International Law of the Sea* (2nd edn, Hart Edition 2016), 3.

[17] Natalie Klein, *Maritime Security and the Law of the Sea* (OUP 2013), 12; Yoshifumi Tanaka, *The International Law of the Sea* (2nd edn, CUP 2015), 17; Donald Rothwell and Tim Stephens, *The International Law of the Sea* (2nd edn, Hart Edition 2016), 2–3; Malcolm Evans, 'Law of the Sea', in Malcolm Evans (ed), *International Law* (4th edn, OUP 2015), 651.

[18] Donald Rothwell and Tim Stephens, *The International Law of the Sea* (2nd edn, Hart Edition 2016), 3; Malcolm Evans, 'Law of the Sea', in Malcolm Evans (ed), *International Law* (4th edn, OUP 2015), 651.

[19] Natalie Klein, *Maritime Security and the Law of the Sea* (OUP 2013), 12; Yoshifumi Tanaka, *The International Law of the Sea* (2nd edn, CUP 2015), 17; Donald Rothwell and Tim Stephens, *The International Law of the Sea* (2nd edn, Hart Edition 2016), 2–4; Malcolm Evans, 'Law of the Sea', in Malcolm Evans (ed), *International Law* (4th edn, OUP 2015), 651.

[20] Donald Rothwell and Tim Stephens, *The International Law of the Sea* (2nd edn, Hart Edition 2016), 4.

[21] Ibid.

[22] Donald Rothwell and Tim Stephens, *The International Law of the Sea* (2nd edn, Hart Edition 2016), 4–5; Malcolm Evans, 'Law of the Sea', in Malcolm Evans (ed), *International Law* (4th edn, OUP 2015), 652; John Noyes, 'The Territorial Sea and Contiguous Zone', in Donald Rothwell et al. (eds), *The Oxford Handbook of the Law of the Sea* (OUP 2015), 93.

law of the sea during the 1930 Hague Conference.[23] Notwithstanding the outbreak of World War II and the hiatus in codification efforts that ensued, repeated claims of zones akin to a territorial sea allowed state practice to thrive.[24]

Between the post-war era and the 1980s, a series of instruments were developed, each dealing with a different topic. All of these ultimately led to the Conferences on the Law of the Sea and resulted in the adoption of the 1982 United Nations Convention on the Law of the Sea, which has been ratified by 168 states.[25] After years of discussions, this quasi-universal legal framework, which can be regarded as the constitution of the seas, now regulates an exceptional range of ocean activities in a holistic and detailed manner.[26]

2.2 International Space Law

Space law emerged as early as 1932, most notably with Vladimir Mandl publishing a monograph on the subject.[27] However, it officially developed in the context of the Cold War and was triggered by one event in particular, the launch of Sputnik 1 by the Soviet Union in 1957, which was the first artificial satellite to be put into the orbit of the Earth.[28] At the time, two competing nations were engaged in the development of space technologies, the United States of America (U.S.) and the Soviet Union.[29]

---

[23] Donald Rothwell and Tim Stephens, *The International Law of the Sea* (2nd edn, Hart Edition 2016), 4–5; Malcolm Evans, 'Law of the Sea', in Malcolm Evans (ed), *International Law* (4th edn, OUP 2015), 652; James Harrison, *Making the Law of the Sea: A Study in the Development of International Law* (CUP 2011), 20.

[24] Donald Rothwell and Tim Stephens, *The International Law of the Sea* (2nd edn, Hart Edition 2016), 4–5.

[25] James Harrison, *Making the Law of the Sea: A Study in the Development of International Law* (CUP 2011), 3; Oceans and Law of the Sea, United Nations, Division for Ocean Affairs and the Law of the Sea, 'Chronological lists of ratifications of, accessions and successions to the Convention and the related Agreements' (3 April 2018) www.un.org/depts/los/reference_files/chronological_l ists_of_ratifications.htm.

[26] James Harrison, *Making the Law of the Sea: A Study in the Development of International Law* (CUP 2011), 27, 48, 51.

[27] Vladimir Kopal, 'Origins of space law and the role of the United Nations', in Christian Brünner and Alexander Soucek, *Outer Space in Society, Politics and Law* (Springer, 2011), 221.

[28] Frans G. von der Dunk, "A Tale of Two Oceans: Governance of Terrestrial and Outer Space 'Global Commons'," *Asian Journal of Air and Space Law* Vol. 2, 1 (2012), 31; M. J. Peterson, 'The Use of Analogies in Developing Outer Space Law International Organization' (1997) 51(2) *International Organization* 245, 245; Nicolas Mateesco Matte, 'The Law of the Sea and Outer Space: A Comparative Survey of Specific Issues' (1982) 3(1) *Ocean Yearbook Online* 13, 14; Jonathan Sydney Koch, 'Institutional Framework for the Province of all Mankind: Lessons from the International Seabed Authority for the Governance of Commercial Space Mining' (2018) 16 *Astropolitics* 1, 2.

[29] Stephan Hobe, 'The Impact of New Developments on International Space Law (New Actors, Commercialisation, Privatisation, Increase in the Number of "Space-faring" Nations)' (2010) 15(4) *Uniform Law Review* 869, 869.

As a result, the laws enacted then still reflect their wishes and compromises, while other nations adopted an observer role.[30]

Between 1967 and 1979, the forum chosen to develop space law, the United Nations Committee on the Peaceful Uses of Outer Space (COPUOS), generated five treaties.[31] Among these, the Outer Space Treaty (OST) is regarded as the nucleus of space law and has been qualified of *quasi*-constitutional or of *Magna Carta* of international space law.[32] To date, it has been ratified by 110 nations.[33] The OST lays the principles that were used as the starting point for the other four treaties.[34] For example, Article VII of the OST contains a general statement about the international liability of states in case of damage. This provision was subsequently used to expand on the subject of liability with the development of the Convention on International Liability for Damage Caused by Space Objects, commonly called the Liability Convention (LIAB). Moreover, the OST has the ambition of "furthering the purposes and principles of the UN Charter" and specifies that space exploration shall

---

[30] Frans G. von der Dunk, 'A Tale of Two Oceans: Governance of Terrestrial and Outer Space "Global Commons"' (2012) 2(1) *Asian Journal of Air and Space Law* 31, 34; Henry Hertzfeld, 'Current and future issues in international space law' (2009*) 15(2) ILSA Journal of International & Comparative Law* 327, 327; Armel Kerrest, 'Space law and the law of the sea', in Christian Brünner and Alexander Soucek, *Outer Space in Society, Politics and Law* (2011, Springer Wien New York), 248.

[31] Treaty on Principles Governing the Activities of States in the Exploration and Use of Outer Space, including the Moon and Other Celestial Bodies (Outer Space Treaty) (opened for signature 27 January 1967, entered into force 10 October 1967) 610 UNTS 8843; Agreement on the Rescue of Astronauts, the Return of Astronauts and Return of Objects Launched into Outer Space (Rescue Agreement) (opened for signature 22 April 1968, entered into force 3 December 1968) 672 UNTS 9574; Convention on International Liability for Damage Caused by Space Objects (Liability Convention) (opened for signature 29 March 1972, entered into force 1 September 1972) 961 UNTS 13810; Convention on The Registration of Objects Launched In Outer Space (Registration Convention) (opened for signature 14 January 1975, entered into force 15 September 1976) 1023 UNTS 15020; Agreement governing the Activities of States on the Moon and Other Celestial Bodies (Moon Agreement) (opened for signature 18 December 1979, entered into force 11 July 1984) 1363 UNTS 23002; Frans Von Der Dunk, 'International Space Law' Frans von der Dunk et al. (eds), *Handbook of Space Law* (Edward Edgar Publishing 2015), 37.

[32] Vladimir Kopal, 'Origins of space law and the role of the United Nations', in Christian Brünner and Alexander Soucek, *Outer Space in Society, Politics and Law* (2011, Springer Wien NewYork), 231; Alexander Soucek, '*International Law*', idem, 298; Joanne I. Gabrynowicz, 'Space Law: Its Cold War Origins and Challenges in the Era of Globalization' (2004), 37 *Suffolk University Law Review*, 1042; Jinyuan Su, 'Legality of unilateral exploitation of space resources under international law' (2017) 66 *International & Comparative Law Quarterly* 991, 993.

[33] Committee on the Peaceful Uses of Outer Space, Legal Subcommittee, 'Status of International Agreements relating to activities in outer space as at 1 January 2020' www.unoosa.org/documents/pdf/spacelaw/treatystatus/TreatiesStatus-2020E.pdf.

[34] Joanne I. Gabrynowicz, 'Space Law: Its Cold War Origins and Challenges in the Era of Globalization' (2004), 37 *Suffolk University Law Review*, 1042.

be conducted in observance of international law.[35] In this way, general international law is incorporated into space law.

Besides treaty law-making, because space activities and the space industry are still relatively nascent, especially compared to maritime activities, no substantial or undisputed international custom has arisen.[36] The principles of the OST, recognized by states, come closest to a custom, largely because there is "no evidence of dissenting practice of non-ratifying states," which proves to be unsatisfactory.[37]

It is this lack of customary international law and the unclarity of existing treaty-law that sparked the writing of this article. Especially since this is mostly the case in regards of highly pressing issues in space law, such as boundary delimitation and flags of convenience. In the face of this vacuum, it appears judicious to appeal to the richness of the law of the sea and its development over millennia, and analyze successful and unsuccessful mechanisms.

3 Analogies

Two significant problems posed by space law and the analogous issues with regard to the law of the sea are presented: delimitation of boundaries and flags of convenience. The aim of the following sections in this part is to determine whether the law of the sea may shed some light on the establishment of new regimes for space law.

3.1 Delimitation of Boundaries

The rise of mega constellations has a direct impact on the outer space delimitation issue. Or rather, the lack of that delimitation will prove to be extremely problematic with the development of mega constellations.

As of January 2021, the European Space Agency's (ESA) account of space debris estimates 34,000 objects over ten centimeters to orbit around Earth, 900,000 between one and ten centimeters, and 128 million between one millimeter and one centimeter.[38] Overall, the millions of pieces of debris come from 10,680 satellites, of which 6,250 are still in space and 3,300 still operational.[39]

---

[35] Outer Space Treaty, Preamble and Article I; Joanne I. Gabrynowicz, 'Space Law: Its Cold War Origins and Challenges in the Era of Globalization' (2004), 37 *Suffolk University Law Review*, 1042.

[36] Peter Malanczuk, 'Space law as a branch of international law' (1994) 25 *Netherlands Yearbook of International Law* 159, 159.

[37] Ibid.

[38] Space Debris User Portal, ESOC, 'Space Environment Statistics' (*ESA*, 29 September 2019) https://sdup.esoc.esa.int/discosweb/statistics/.

[39] For information on the difference between debris and active satellites see Ewan Wright, 'Legal aspects relating to satellite constellations' in: Annette Froehlich (ed) *Legal Aspects Around Satellite*

A mere 28,200 objects are actually traced and catalogued.[40] Amongst those, over 90% are non-operational and around 15,500 are in LEO.[41] Debris mostly come from payload and payload fragmentation debris, followed by rocket fragmentation debris, unidentified objects, and rocket bodies.[42] LEO is thus the most populated, with concurrently the highest number of debris and of operational systems, i.e. between 70 and 75%.[43] It makes it the region most vulnerable to collisions.[44] The gravity of those collisions is itself exacerbated, for incidents in LEO are "400 times more destructive than in geostationary orbit (GEO) because of the higher orbital velocity of objects and the greater inclinations."[45]

Over 40 large objects reenter the atmosphere each year, objects of one meter or above do so each week, and about two small *tracked* objects each day.[46] If small objects burn upon reentry, larger ones and those made of high-melting-point materials reach Earth's surface.[47] When they do, a procedure of bringing them to a relatively unpopulated area, often in the ocean, exists so as to lower the risk of harming someone.[48] Putting aside the oceanic pollution caused by this practice, not all large objects are able to control their reentry.[49] Also, solar effects on the atmosphere, aerodynamics and dynamics condition the entry location and makes the exact entry point

---

Constellations (Springer 2019); Space Debris User Portal, ESOC, 'Space Environment Statistics' (*ESA*, 29 September 2019) https://sdup.esoc.esa.int/discosweb/statistics/.

[40]Space Debris User Portal, ESOC, 'Space Environment Statistics' (*ESA*, 29 September 2019) https://sdup.esoc.esa.int/discosweb/statistics/.

[41]Space Debris User Portal, ESOC, 'Space Environment Statistics' (*ESA*, 29 September 2019) https://sdup.esoc.esa.int/discosweb/statistics/; European Space Imaging, 'The Lifespan of Orbiting Satellites' (*ESI*, 11 March 2019) www.euspaceimaging.com/the-lifespan-of-orbiting-satellites/; Nathan Reiland et al., 'The Dynamical Placement of Mega-Constellations' (42nd COSPAR Scientific Assembly, Pasadena, July 2018).

[42]Ibid.

[43]Brian Patrick Hardy, 'Long-term effects of satellite megaconstellations on the debris environment in low earth orbit' (M.S. thesis, University of Illinois, 11 May 2020), 1–2; Martyn Warwick, 'Already dangerous, satellite dodgems gets more perilous by the day' (*Telecom TV*, 3 August 2020) www.telecomtv.com/content/access-evolution/already-dangerous-satellite-dodgems-gets-more-perilous-by-the-day-39362/.

[44]Brian Patrick Hardy, 'Long-term effects of satellite megaconstellations on the debris environment in low earth orbit' (M.S. thesis, University of Illinois, 11 May 2020), 1–2.

[45]Philip Chrystal et al., 'Space Debris: On Collision course for insurers?' (Swiss Re, 2011); William Ailor, 'Hazards of Reentry Disposal of Satellites from Large Constellations' (COPUOS Scientific Subcommittee, Vienna, 2020).

[46]ESA, 'Reentry and collision avoidance' (ESA, n.d.) www.esa.int/Safety_Security/Space_Debris/Reentry_and_collision_avoidance; William Ailor, 'Space Debris Reentry Hazards' (COPUOS Scientific Subcommittee, Vienna, 2012).

[47]National Research Council, 'Limiting Future Collision Risk to Spacecraft: An Assessment of NASA's Meteoroid and Orbital Debris Programs' (The National Academies Press, 2011), 60.

[48]European Space Imaging, 'The Lifespan of Orbiting Satellites' (*ESI*, 11 March 2019) www.euspaceimaging.com/the-lifespan-of-orbiting-satellites/.

[49]William Ailor, 'Space Debris Reentry Hazards' (COPUOS Scientific Subcommittee, Vienna, 2012).

unpredictable.[50] Especially because a satellite reentering the atmosphere is going to generate a number of fragments after the reentry breakup process. The cloud of surviving debris can spread over 2000 km of ground.[51]

The problem with objects surviving reentry when it comes to delimiting airspace and outer space is not where those objects end up on the surface but what they may come in contact with on their way there. That is especially the case if the event occurs at a "contested" altitude, one that could be considered both airspace and outer space. The number of reentries and uncontrolled ones is increasing.[52] William Ailor determined in a 2019 study on the "hazards of reentry disposal of satellites from large constellations", that in 2030 "the probability of debris striking a commercial aircraft would be 0.001/year" (1 per 1000 years) and casualties for people in aircrafts could reach to 0.3/year (3 per 10 years) "without emergency actions [taken] by pilots."[53] That only reflects commercial air traffic. If all global flights were to be accounted for, the odds would be even higher. Also, the research is based on around 16,000 satellites shared between different constellations and assumes all reentries follow natural orbital decay.

Nevertheless, according to Analytical Graphics, at least 50,000 new systems will be launched in the near future.[54] Current mega constellation plans for Starlink from SpaceX and Kuiper from Amazon could, between them, add up to 45,000 small satellites in orbit if approved by the U.S. Federal Communications Commission (FCC).[55] Starlink alone will double debris in LEO if even "the minimum regulatory requirements for post-mission disposal is met."[56]

However, post-mission may come sooner than expected. The usual operational lifetime of satellites—different from its orbital lifetime—ranges between several weeks for the smaller systems to about 20 years for the larger ones.[57] Considering that, it was calculated in 2010 that "large satellites in LEO can be expected to incur a mean lifetime reduction of around 13% […] with much of this reduction coming

---

[50] Ibid.

[51] William Ailor, 'Large Constellation Disposal Hazards' (*Center for Space Policy and Strategy*, January 2020), 3.

[52] Jarbas Cordeiro Sampaio, Ewerton Felipe B. P. dos Santos, 'Space Debris: Reentry and Collision Risk' (2020) 7 *Proceeding Series of the Brazilian Society of Computational and Applied Mathematics* 1.

[53] William Ailor, 'Hazards of reentry disposal of satellites from large constellations' (2019) 6 Journal of Space Safety Engineering 2, 113–121, 113.

[54] Martyn Warwick, 'Already dangerous, satellite dodgems gets more perilous by the day' (*Telecom TV*, 3 August 2020) www.telecomtv.com/content/access-evolution/already-dangerous-satellite-dodgems-gets-more-perilous-by-the-day-39362/.

[55] Brian Patrick Hardy, 'Long-term effects of satellite megaconstellations on the debris environment in low earth orbit' (M.S. thesis, University of Illinois, 11 May 2020), ii.

[56] Ibid.

[57] George Fox et al., 'A Satellite Mortality Study to Support Space Systems Lifetime Prediction' (IEEE Aerospace Conference, Big Sky, March 2013).

from degradation of solar panels by small un-trackable debris."[58] Since then, the proliferation of mega constellation plans could double the 2010 results and in some cases the reduction could reach 60%; a large LEO satellite supposed to have a 20-year-lifetime could last eight years instead.[59]

In addition to this, a research published in 2020 demonstrated that current mega constellation designs are sub-optimal when it comes to collision assessment and minimization; operators do not seem to use astrodynamics tools that have the potential to tackle space debris at conception before they become a problem.[60]

All those elements demonstrate that mega constellations increase collisions risks in airspace and in zones that could presently be considered both airspace and outer space. The same can be said from risks between objects of different natures, i.e. aeronautic objects and space objects. Statistics should at least ring the alarm and trigger concrete decision-making on the delimitation between airspace and outer space and as a consequence the election of the legal regime.

Against this backdrop, many states have been reluctant to agree on a universal definition of outer space and thus on a definition of space law. If no agreement has emerged from decades of debates and legal writings, the topic nonetheless remains on the agenda of the COPUOS Working Group.[61] A number of countries have indicated their wish to see a boundary and a clear international legal regime set.[62] However,

---

[58] William Ailor, 'Effect of large constellations on lifetime of satellites in low earth orbits' (2017) 4 *Journal of Space Safety Engineering* 3–4, 117–123, 117.

[59] Ibid.

[60] Nathan Reiland et al., 'Assessing and Minimizing Collisions in Satellite Mega-Constellations' (Advanced Maui Optical and Space Surveillance Technologies Conference, Maui, 17–20 September 2019), 23.

[61] COPUOS 2019, Working Groups of the Committee and its Subcommittees (UNOOSA) www.unoosa.org/oosa/en/ourwork/copuos/working-groups.html#LSCWGDD; COPUOS, Working Group on the Definition and Delimitation of Outer Space of the Legal Subcommittee (UNOOSA) www.unoosa.org/oosa/en/ourwork/copuos/lsc/ddos/index.html; Olavo de Oliveira Bittencourt Neto, 'Delimitation of outer space and Earth orbits', in Yanal Abul Failat and Anél Ferreira-Snyman (eds), *Outer Space Law: Legal Policy and Practice* (Globe Law and Business Ltd 2017), 55.

[62] E.g. Saudi Arabia in UNGA 'Definition and delimitation of outer space: views of States members and permanent observers of the Committee' Note by the Secretariat, Addendum (11 January 2019) COPUOS LSC, UN Doc A/AC.105/1112/Add.6; Tunisia in UNGA, 'Information relating to any practical case known that would warrant the definition and delimitation of outer space' Note by the Secretariat (24 January 2020) COPUOS LSC, UN Doc A/AC.105/1226; Turkey in UNGA, 'Definition and delimitation of outer space: views of States members and permanent observers of the Committee' Note by the Secretariat (28 January 2016) COPUOS LSC, UN Doc A/AC.105/1112 and UNGA, 'Definition and delimitation of outer space: views of States members and permanent observers of the Committee' Note by the Secretariat, Addendum (18 January 2017) COPUOS LSC, UN Doc A/AC.105/1112/Add.2; South Africa in UNGA 'Definition and delimitation of outer space: views of States members and permanent observers of the Committee' Note by the Secretariat, Addendum (22 January 2018) COPUOS LSC, A/AC.105/1112/Add.4; Mexico in UNGA, 'Definition and delimitation of outer space: views of States members and permanent observers of the Committee' Note by the Secretariat (12 February 2016), COPUOS LSC, UN Doc A/AC.105/1112/Add.1; Myanmar and Pakistan in UNGA, 'Questions on suborbital flights for scientific missions and/or for human transportation' Note by the Secretariat, Addendum (11 January 2019) COPUOS LSC, UN Doc A/AC.105/1039/Add.12; Peru in UNGA, 'Questions on

their views manifestly do not weight as much as the ones of space powers, since no mechanism has been put into motion.

By comparison, agreeing on the maritime zone system as we know it was the result of lengthy discussions and incremental adaptations, notably within the International Law Commission (ILC) and subsequently during the Conferences on the Law of the Sea.[63] The ILC, like COPUOS, established the importance of settling the delimitation matter early on.[64] And exactly as is the case for space law today, the law of the sea then had to start building the maritime zone system with scarce resources, as not much practice or jurisprudence existed.[65] Fortunately for space law, maritime zones and the territorial sea-high seas boundary offer a parallel to inspire understanding of the airspace-outer space boundary.[66]

3.1.1 The Airspace-Outer Space Boundary Debate

The current legal landscape of the airspace-outer space delimitation is extremely unclear.[67] Both Tanaka and Rothwell have referred to the crucial importance in the law of the sea to settle the extent of coastal states' jurisdiction and as such the physical limits of national sovereignty.[68] Equivalently, a primordial task for space law is to settle the height at which states may no longer claim territorial sovereignty. In other words, it should determine where its application starts, and by consequence where the application of air law shall end.[69] In legal fields such as the law of the sea and space law, the *locus actus* is determinant for it dictates the applicable legal regime.

---

suborbital flights for scientific missions and/or for human transportation' Note by the Secretariat Addendum (21 January 2020) COPUOS LSC, UN Doc A/AC.105/1039/Add.13; WHO in UNGA, 'Definition and delimitation of outer space: views of States members and permanent observers of the Committee' Note by the Secretariat, Addendum (18 January 2017) COPUOS LSC, UN Doc A/AC.105/1112/Add.2.

[63] Donald Rothwell and Tim Stephens, *The International Law of the Sea* (2nd edn, Hart Edition 2016), 415.

[64] He Qizhi, 'The Problem of Definition and Delimitation of Outer Space' (1982) 10 *Journal of Space Law* 157, 157; Committee on the Peaceful uses of Outer Space, Legal Sub-committee, 'The question of the definition and/or the delimitation of outer space' background paper prepared by the Secretariat in 1970 and updated in 1977, A/AC.105/C.2/7 (7 May 1970) www.unoosa.org/pdf/limited/c2/AC105_C2_L007E.pdf.

[65] Donald Rothwell and Tim Stephens, *The International Law of the Sea* (2nd edn, Hart Edition 2016), 415.

[66] Frans G. von der Dunk, 'The Sky Is the Limit - But Where Does It End? New Developments on the Issue of Delimitation of Outer Space' (2005) *Proceedings of the 48th Colloquium on the Law of Outer Space* 84, 85; Francis Lyall and Paul Larsen, *Space Law: A Treatise* (2nd edn, Taylor & Francis 2017), 135.

[67] Mark J. Sundahl, 'Legal status of spacecraft', in Ram S. Jakhu and Paul Stephen Dempsey (eds), *Routledge Handbook of Space Law* (Routledge Handbooks 2016), 53.

[68] Yoshifumi Tanaka, *The International Law of the Sea* (2nd edn, CUP 2015), 44; Donald Rothwell and Tim Stephens, *The International Law of the Sea* (2nd edn, Hart Edition 2016), 75.

[69] Jinyuan Su, 'The Delineation Between Airspace and Outer Space and the Emergence of Aerospace Objects' (2013) 78 *Journal of Air Law and Commerce* 355, 357.

However, whereas the law of the sea has now set maritime zones, the lower limit of outer space has still not been settled, either scientifically or legally.[70] As such, arguing that space law applies in outer space without having defined "outer space" is devoid of sense. According to this reasoning, proclaiming that space law applies to space activity is of no help either.

Although states were provided with a forum for discussion within the legal subcommittee of COPUOS and various questionnaires, most have refrained from sharing their opinion on the boundary.[71] While more are speaking, they often disagree not only on the process to determine the boundary, but also on whether a delimitation is needed in the first place.

On the latter, statements such as the one made by Viet Nam at COPUOS in January 2020 according to which no case can be identified as warranting the definition of outer space and its boundary, conveniently focus on the past rather than on the foreseeable future.[72] The research and numbers laid above, as well as the official plans of mega constellations are nothing but evidence that a limit has to be set. There also seem to be a correlation between countries against the boundaries and countries with unreasonable assertions of sovereignty over airspace. In the case of Viet Nam, according to a 2002 Decision taken by the Ministry of National Defence, the national airspace over Hanoi goes to infinity.[73]

It has also been contended that there is no need for a demarcation because airspace and outer space lack a clear, definite, physical line.[74] This is debatable. Indeed, it all depends on the type of line that is being explored, the atmospheric layers for instance have some sort of separations. Nonetheless, although a legal demarcation should remain as close as possible to physical realities, it may depart from it.[75] Therefore, the lack of a physical line does not prevent establishing a line of another

---

[70] Mark J. Sundahl, 'Legal status of spacecraft', in Ram S. Jakhu and Paul Stephen Dempsey (eds), *Routledge Handbook of Space Law* (Routledge Handbooks 2016), 53.

[71] Jinyuan Su, 'The Delineation Between Airspace and Outer Space and the Emergence of Aerospace Objects' (2013) 78 *Journal of Air Law and Commerce* 355, 362; Committee on the Peaceful Uses of Outer Space, Legal Subcommittee, 41st session, 'Historical summary on the consideration of the question on the definition and delimitation of outer space' A/AC.105/769 (18 January 2002) www.unoosa.org/pdf/reports/ac105/AC105_769E.pdf.

[72] UNGA, 'Information relating to any practical case known that would warrant the definition and delimitation of outer space' Note by the Secretariat (24 January 2020) COPUOS LSC, UN Doc A/AC.105/1226.

[73] Decision no. 11/2004/QD-BQP (30 October 2002); UNGA, 'National legislation and practice relating to the definition and delimitation of outer space' Note by the Secretariat Addendum (20 January 2020) COPUOS LSC, UN Doc A/AC.105/865/Add.23.

[74] Dean N Reinhardt, 'The Vertical Limit of State Sovereignty' (2007) 72 *Journal of Air Law and Commerce* 65, 113; Olavo de Oliveira Bittencourt Neto, 'Delimitation of outer space and Earth orbits', in Yanal Abul Failat and Anél Ferreira-Snyman (eds), *Outer Space Law: Legal Policy and Practice* (Globe Law and Business Ltd 2017), 53.

[75] Olavo de Oliveira Bittencourt Neto, 'Delimitation of outer space and Earth orbits', in Yanal Abul Failat and Anél Ferreira-Snyman (eds), *Outer Space Law: Legal Policy and Practice* (Globe Law and Business Ltd 2017), 59; Francis Lyall and Paul Larsen, *Space Law: A Treatise* (2nd edn, Taylor & Francis 2017), 150; *contra* Committee on the Peaceful Uses of Outer Space, Legal Subcommittee, 41st session, 'Historical summary on the consideration of the question on the definition and

nature. For instance, maritime zones, despite having no actual physical demarcation, have been set in a "mathematical" but flexible manner.

Acknowledging that an airspace-outer space boundary is definitely needed, the two main grounds for its determination, functional and spatial, are discussed below.[76]

Functional Approach

The functional theory holds that it is the activity, its objective and purpose, and not the *locus* of the craft, that determines the applicable regime.[77] Aeronautical activities or engines fall under air law and astronautical activities or engines under space law, no matter their altitude.[78] However, this reasoning presents a significant flaw, in that it presupposes that activities or engines can be clearly defined as either aeronautic or astronautic. Indeed, it does not account for recent activities that blur the line between the two, such as suborbital flights and aerospace crafts in general, which can operate in both airspace and outer space.[79] Perhaps the functionalist approach can solve the

---

delimitation of outer space' A/AC.105/769 (18 January 2002) www.unoosa.org/pdf/reports/ac105/AC105_769E.pdf, paras 4, 5.

[76] Katherine M. Gorove, 'Delimitation of Outerspace and the Aerospace Object - Where is the Law' (2000) 28 *Journal of Space Law* 11, 16; Bin Cheng, 'The Legal Status of Outer Space and Relevant Issues: Delimitation of Outer Space and Definition of Peaceful Use' (1983) 11 *Journal of Space Law* 89, 93; Theodore W. Goodman, 'To the End of the Earth: A Study of the Boundary between Earth and Space' (2010) 36 *Journal of Space Law* 87, 89.

[77] Jinyuan Su, 'The Delineation Between Airspace and Outer Space and the Emergence of Aerospace Objects' (2013) 78 Journal of Air Law and Commerce 355, 363; Marietta Benkö and Engelbert Plescher, Space Law: Reconsidering the Definition/Delimitation Question and the Passage of Spacecraft through Foreign Airspace (Eleven International Publishing 2013), 35; Stephen E. Doyle, 'A concise history of space law' (2010) International Institute of Space Law 1,1, www.iislweb.org/website/docs/2010keynote.pdf; Mark J. Sundahl, 'Legal status of spacecraft', in Ram S. Jakhu and Paul Stephen Dempsey (eds), Routledge Handbook of Space Law (Routledge Handbooks 2016), 55.

[78] Marietta Benkö and Engelbert Plescher, *Space Law: Reconsidering the Definition/Delimitation Question and the Passage of Spacecraft through Foreign Airspace* (Eleven International Publishing 2013), 35; Stephen E. Doyle, 'A concise history of space law' (2010) *International Institute of Space Law* 1, 2; Frans G. von der Dunk, 'The Sky Is the Limit - But Where Does It End? New Developments on the Issue of Delimitation of Outer Space' (2005) *Proceedings of the 48th Colloquium on the Law of Outer Space* 84, 85; Mark J. Sundahl, 'Legal status of spacecraft', in Ram S. Jakhu and Paul Stephen Dempsey (eds), *Routledge Handbook of Space Law* (Routledge Handbooks 2016), 55; Francis Lyall and Paul Larsen, *Space Law: A Treatise* (2nd edn, Taylor & Francis 2017), 150.

[79] Stephen E. Doyle, 'A concise history of space law' (2010) *International Institute of Space Law* 1, 2, www.iislweb.org/website/docs/2010keynote.pdf; Mark J. Sundahl, 'Legal status of spacecraft', in Ram S. Jakhu and Paul Stephen Dempsey (eds), *Routledge Handbook of Space Law* (Routledge Handbooks 2016), 53, 55; Olavo de Oliveira Bittencourt Neto, 'Delimitation of outer space and Earth orbits', in Yanal Abul Failat and Anél Ferreira-Snyman (eds), *Outer Space Law: Legal Policy and Practice* (Globe Law and Business Ltd 2017), 55–56; Francis Lyall and Paul Larsen, *Space Law: A Treatise* (2nd edn, Taylor & Francis 2017), 150.

problem in relation to clear-cut technologies, but ultimately it remains unhelpful for settling the regime applicable to more complex technologies.[80]

Su has sought to counter this argument by stating that "aerospace objects performing Earth-to-space transportation are space objects governed by space law, while aerospace objects performing Earth-to-Earth transportation are governed by air law even though they may temporarily traverse outer space."[81] However, this is unsatisfactory and devoid of logic. Indeed, the origin of the problem is precisely that there is no agreed legal definition of where Earth ends and where outer space begins. Therefore, there is no definition of what is "Earth" and what is "space" for the purposes of the law. As a consequence, how should one distinguish Earth-to-space travel from Earth-to-Earth travel?

A valid point was made by the astrophysicist McDowell, namely that applying one regime rather than another to a vehicle, according to its purpose instead of its location, leads to untenable situations, for example, when a vehicle governed by space law collides or interferes with a vehicle governed by air law.[82] In other words, when two such vehicles and the opposite laws regulating them come into conflict, who is to decide on the applicable legal regime and how?[83]

Lastly, because the functional approach is based on determining the nature of activities, a rather pragmatic argument can be made against it. It seems doubtful that states would reveal with full transparency the nature of their activities, which most often have military implications. Even if they were to willingly reveal all the ramifications of their activities, some of those may in fact serve multiple purposes. Which regime shall then be assigned?

Spatial Approach

The spatial approach seeks to delimit airspace and outer space on the basis of a vertical limit. Yet, the *modus operandi* for choosing this height varies.[84] The Karman line, or

---

[80] Stephen E. Doyle, 'A concise history of space law' (2010) *International Institute of Space Law* 1, 2, www.iislweb.org/website/docs/2010keynote.pdf.

[81] Jinyuan Su, 'The Delineation Between Airspace and Outer Space and the Emergence of Aerospace Objects' (2013) 78 *Journal of Air Law and Commerce* 355, 369.

[82] Jonathan C. McDowell, 'The edge of space: Revisiting the Karman Line' (2018) 151 *Acta Astronautica* 668, 668.

[83] Bin Cheng, 'The Legal Status of Outer Space and Relevant Issues: Delimitation of Outer Space and Definition of Peaceful Use' (1983) 11 *Journal of Space Law* 89, 93.

[84] Bin Cheng, *Studies in International Space Law* (OUP 1997), 426; Jinyuan Su, 'The Delineation Between Airspace and Outer Space and the Emergence of Aerospace Objects' (2013) 78 *Journal of Air Law and Commerce* 355, 363; Marietta Benkö and Engelbert Plescher, *Space Law: Reconsidering the Definition/Delimitation Question and the Passage of Spacecraft through Foreign Airspace* (Eleven International Publishing 2013), 31; Stephen E. Doyle, 'A concise history of space law' (2010) *International Institute of Space Law* 1, 1, www.iislweb.org/website/docs/2010keynote.pdf.

the aerodynamic lift theory, and the lowest point of orbital flight, also called lowest perigee demarcation, have been predominantly put forward.[85]

First, the Karman line refers to the point beyond which aerodynamic lift gives in to centrifugal force.[86] In other words, aerodynamic lift, which essentially refers to the force that lifts an object through a fluid, in this case air, is taken over by centrifugal force which, simply put, is a force that acts on an object by pushing it away from a center. Either the centripetal force, which pushes an object towards a center, is equal to the centrifugal force, and the object moves in orbit, or it is inferior, and the object flies away, off the orbit. Whilst the Karman line is commonly thought to be around 100 km from sea level, this has recently been scientifically demonstrated to be incorrect.[87] Apparently, this line lies at around 80 km.[88] It is yet to be seen whether states that used the 100 km height on the basis of the Karman line will now change their practice.

Second, the lowest point of orbital flight basically refers to the reverse phenomenon. This is the lowest altitude a satellite can reach before centrifugal force is no longer able to maintain it in orbit and it falls back towards Earth.[89] According to Perek, this reasoning should determine the boundary, because artificial satellites of the Earth could then be said to move in outer space.[90] However, this proposal has a downside. It is based on Earth's satellites and omits vehicles not operating on its orbits such as uborbital flights or reusable rockets which land back on Earth. For instance, when SpaceX launched Falcon Heavy, after having completed their task the boosters were shut down, stage separation occurred at around 60 km, from where the boosters continued upwards to approximately 90–110 km, before coming back and landing on the ground.[91] In a parallel argument, neither does the lowest point of

---

[85] Mark J. Sundahl, 'Legal status of spacecraft', in Ram S. Jakhu and Paul Stephen Dempsey (eds), *Routledge Handbook of Space Law* (Routledge Handbooks 2016), 55.

[86] Theodore W. Goodman, 'To the End of the Earth: A Study of the Boundary between Earth and Space' (2010) 36 *Journal of Space Law* 87, 91.

[87] Jonathan C. McDowell, 'The edge of space: Revisiting the Karman Line' (2018) 151 *Acta Astronautica* 668.

[88] Ibid., 668, 668.

[89] D. Goedhuis, 'Some Observations on the Problem of the Definition and/or the Delimitation of Outer Space' (1977) 2 *Annals of Air & Space Law* 302; Marietta Benkö and Engelbert Plescher, *Space Law: Reconsidering the Definition/Delimitation Question and the Passage of Spacecraft through Foreign Airspace* (Eleven International Publishing 2013), 32.

[90] Lubos Perek, 'Scientific Criteria for the Delimitation of Outer Space' (1977) 5 *Journal of Space Law* 111, 118; He Qizhi, 'The Problem of Definition and Delimitation of Outer Space' (1982) 10 *Journal of Space Law* 157, 160.

[91] Tariq Malik, 'Success! SpaceX Launches Falcon Heavy Rocket on Historic Maiden Voyage' (Space.com, 6 February 2018) www.space.com/39607-spacex-falcon-heavy-first-test-flight-launch.html; Doris Elin Salazar, 'SpaceX's Falcon Heavy Rocket: By the Numbers' (Space.com, 6 February 2018) www.space.com/39603-spacex-falcon-heavy-rocket-by-the-numbers.html; SpaceX, Falcon Heavy Test Flight (YouTube, 6 February 2018) www.youtube.com/watch?v=wbSwFU6tY1c&t=919s; Harry Pettit and Cheyenne Macdonald, 'The galaxy's fastest car: Elon Musk releases stunning real-time video of Tesla Roadster and its passenger zooming through space' (Dailymail, 7

orbital flight theory take into account objects such as Voyager 1 and 2 and SpaceX's "Starman", which will never enter Earth's atmosphere again.[92]

These examples indubitably all constitute space activities, no matter the altitude of their occurrence. What legal regime should then be applied? In fact, Falcon Heavy's maiden launch provides a prime example. When the boosters headed back towards the ground, what would have happened, if at a 50 km altitude, one of the boosters collided with an aircraft, either because it lost control or because the aircraft was flying on an unexpected or unauthorized path? On the one hand, the booster is a spacecraft and its operation constitutes a space activity. According to the functionalist view, it should therefore be governed by space law. On the other hand, the aircraft falls under the regime of air law. This illustrates the point made by McDowell earlier. However, if the spatial approach is adopted and a boundary is set at a specific height, everything occurring under this limit, such as the imagined collision between a booster and an aircraft, would necessarily have to comply with air law.[93] No place is therefore left for equivocacy on the applicable law. This demonstrates that, contrary to the claims of some academics, such as King, the spatial approach leads to a consistent and unambiguous framework for aerospace objects.[94]

To conclude the debate on the two approaches, if Oduntan has stressed that neither offers satisfactory answers to the delimitation question and that criticism will follow whatever perspective is taken, the spatial approach should be favored.[95] However, it is true that none of the sub-theories proposed by the spatial approach seem adequate to determine a height. A solution may be found by looking at how the law of the sea determined the breadth of the territorial sea and the configurations of its outer limit.

3.1.2 Maritime Zones

It should be expressed that issues of delimitation between adjacent or opposing states are not the subject of this section. First, it is demonstrated that the delimitation between the territorial sea and the high seas resembles the most the delimitation between airspace and outer space and its implications. If other maritime zones exist, examining them would serve no purpose that would advance the present problematic.

February 2018) www.dailymail.co.uk/sciencetech/article-5361789/Ride-Elon-Musks-Starman-travels-space.html.

[92] Jet Propulsion Laboratory, Voyager Mission Status (NASA) https://voyager.jpl.nasa.gov/mission/status/; Mike Wall, 'SpaceX's 'Starman' and Its Tesla Roadster Are Now Beyond Mars' (Space.com, 3 November 2018) www.space.com/42337-spacex-tesla-roadster-starman-beyond-mars.html.

[93] Mark J. Sundahl, 'Legal status of spacecraft', in Ram S. Jakhu and Paul Stephen Dempsey (eds), *Routledge Handbook of Space Law* (Routledge Handbooks 2016), 55.

[94] Matthew T. King, 'Sovereignty's Gray Area: The Delimitation of Air and Space in the Context of Aerospace Vehicles and the Use of Force' (2016) 81 *Journal of Air Law and Commerce* 377, 432; Stephen E. Doyle, 'A concise history of space law' (2010) *International Institute of Space Law* 1, 1 www.iislweb.org/website/docs/2010keynote.pdf.

[95] Gbenga Oduntan, 'The Never Ending Dispute: Legal Theories on the Spatial Demarcation Boundary Plane between Airspace and Outer Space' (2003) 1(2) *Hertfordshire Law Journal* 64, 81.

Also, a part of the sea where the territorial sea is directly adjacent to the high seas is perfectly imaginable. This is the case if the coastal state has not claimed an exclusive economic zone (EEZ) or a contiguous zone. Second, the process of establishing the outer limit of the territorial sea will be examined.

Territorial Sea Versus the High Seas: The Absolute Jurisdiction of the Coastal State Versus the Exclusive Jurisdiction of the Flag State

The airspace-outer space boundary issue resembles the opposition of regimes between the territorial seas and the high seas. On one hand, the territorial sea and the air space above it are all subject to national sovereignty.[96]

On the other hand, the high seas and outer space are international territories within which no state shall make claims of sovereignty.[97] On the high seas, the exclusive jurisdiction of the flag state applies.[98] In outer space, the same applies to the state of registry.[99] It is critical to evaluate where this change of regime occurs. With regard to the sea, this has an impact on what a state can and cannot do, and in space it additionally affects the applicable branch of the law.

Breadth of the Territorial Sea and Baselines

To begin, according to Article 3 of the UNCLOS, a state *may* claim a territorial sea, the breadth of which shall not exceed 12 nm. Arguably, before this provision, the standard practice was a breadth of 3 nm, which was as far as a cannon could shoot.[100] Today, the majority of states abide by Article 3, claiming a territorial sea of 12 nm.[101]

---

[96] UNCLOS, art 2; James Kraska, *Maritime Power and the Law of the Sea: Expeditionary Operations in World Politics* (OUP 2011), 109; Yoshifumi Tanaka, *The International Law of the Sea* (2nd edn, CUP 2015), 84–86; H. A. Wassenbergh, 'Parallels and Differences in the Development of Air, Sea and Space Law in the Light of Grotius' (1984) 9 *Annals of Air Space Law* 163, 173; Nicolas Mateesco Matte, 'The Law of the Sea and Outer Space: A Comparative Survey of Specific Issues' (1982) 3(1) *Ocean Yearbook Online* 13, 20–21.

[97] William V. Dunlap, 'International Boundaries: The Next Generation' (1999–2000) *IBRU Boundary and Security Bulletin* 106, 107, 109; Nicolas Mateesco Matte, 'The Law of the Sea and Outer Space: A Comparative Survey of Specific Issues' (1982) 3(1) *Ocean Yearbook Online* 13, 21.

[98] UNCLOS, art 94(1).

[99] Outer Space Treaty, art VIII.

[100] James Kraska, *Maritime Power and the Law of the Sea: Expeditionary Operations in World Politics* (OUP 2011), 114–116; David Harris and Sandesh Sivakumaran, *Cases and Materials on International Law* (8th edn, Sweet & Maxwell 2015), 328; John Noyes, 'The Territorial Sea and Contiguous Zone', in Donald Rothwell et al. (eds), *The Oxford Handbook of the Law of the Sea* (OUP 2015), 92–93.

[101] Donald Rothwell and Tim Stephens, *The International Law of the Sea* (2nd edn, Hart Edition 2016), 73–74; David Harris and Sandesh Sivakumaran, *Cases and Materials on International Law* (8th edn, Sweet & Maxwell 2015), 328; Dean N. Reinhardt, 'The Vertical Limit of State

A few states have made controversial and excessive claims, but these claims have not been embraced by the international community.[102]

Agreeing on the breadth of the territorial sea was merely the beginning. Another issue was to determine where the 12 nm would commence. The normal rule requires following the low-water mark of the shoreline, in terms of which every point is at the same distance from its "twin point" situated on the outer limit.[103] In other words, if a state claims a territorial sea of 12 nm, then every point of the outer line will be at 12 nm from the baseline, therefore mimicking the structure of the latter.

Having said that, there is an exception to this rule. Upon meeting certain exceptional circumstances, for instance when the coastline is too indented, coastal states may use the straight baseline method.[104] Notwithstanding its fictional nature, its trajectory must remain reasonably and coherently linked to the general shape of the coastline.[105] Interestingly, the use of straight baselines, before being posited in Article 5 of the UNCLOS, was asserted in the *Anglo-Norwegian Fisheries Case*.[106] Therein, Norway argued that it was entitled to use straight baselines with regard to geographical characteristics such as deep indentations caused by fjords, but this was contested by the UK.[107] Ultimately, the Court found that the method used by Norway

---

Sovereignty' (2007) 72 *Journal of Air Law and Commerce* 65, 131; John Noyes, 'The Territorial Sea and Contiguous Zone', in Donald Rothwell et al. (eds), *The Oxford Handbook of the Law of the Sea* (OUP 2015), 94–95.

[102] Yoshifumi Tanaka, *The International Law of the Sea* (2nd edn, CUP 2015), 85; Donald Rothwell and Tim Stephens, *The International Law of the Sea* (2nd edn, Hart Edition 2016), 73, 75; David Harris and Sandesh Sivakumaran, *Cases and Materials on International Law* (8th edn, Sweet & Maxwell 2015), 328.

[103] UNCLOS, arts 4, 5; *Anglo-Norwegian Fisheries Case*, Judgment, ICJ Reports 1951, p 128; James Kraska, *Maritime Power and the Law of the Sea: Expeditionary Operations in World Politics* (OUP 2011), 109; Yoshifumi Tanaka, *The International Law of the Sea* (2nd edn, CUP 2015), 45; David Harris and Sandesh Sivakumaran, *Cases and Materials on International Law* (8th edn, Sweet & Maxwell 2015), 329; Malcolm Evans, 'Law of the Sea', in Malcolm Evans (ed), *International Law* (4th edn, OUP 2015), 654.

[104] UNCLOS, art 7; *Maritime Delimitation and Territorial Questions between Qatar and Bahrain* (Merits), Judgment, ICJ Reports 2001, para 212; James Kraska, Maritime Power and the Law of the Sea: Expeditionary Operations in World Politics (OUP 2011), 110; Yoshifumi Tanaka, The International Law of the Sea (2nd edn, CUP 2015), 47; Donald Rothwell and Tim Stephens, The International Law of the Sea (2nd edn, Hart Edition 2016), 5–6; Malcolm Evans, 'Law of the Sea', in Malcolm Evans (ed), International Law (4th edn, OUP 2015) 654–655.

[105] UNCLOS, art 7(3); *Anglo-Norwegian Fisheries Case*, Judgment, ICJ Reports 1951, p 129–130; David Harris and Sandesh Sivakumaran, *Cases and Materials on International Law* (8th edn, Sweet & Maxwell 2015), 330.

[106] Yoshifumi Tanaka, *The International Law of the Sea* (2nd edn, CUP 2015), 48.

[107] David Harris and Sandesh Sivakumaran, *Cases and Materials on International Law* (8th edn, Sweet & Maxwell 2015), 329; Malcolm Evans, 'Law of the Sea', in Malcolm Evans (ed), *International Law* (4th edn, OUP 2015), 655.

was in accordance with international law.[108] Later, most of the principles sustained in the case were reaffirmed in the UNCLOS.[109]

Regrettably, the settlement of baselines has often been abused, especially because of the misuse of the straight baseline method.[110] Although the latter should remain an exception to the rule, more than 60 coastal states have attested to these exceptional criteria.[111] The potential cause is the large margin for interpretation left by the UNCLOS, without any safeguard to frame extravagant claims.[112] States without legitimate claims have attempted to use the rule in order to lay claim to as much territory as possible, and therefore to extend their sovereignty and jurisdiction.[113]

3.1.3 Replicating the Territorial Sea-High Seas Boundary

The previous section demonstrated that vessels at sea are subject to different regimes depending on the maritime zone they navigate through. A vessel sailing within a coastal state's territorial sea is subject to its domestic laws, jurisdiction and sovereignty. Once the vessel reaches the high seas, it is subjected to the exclusive jurisdiction of the flag state and has to abide by international standards. This proves that setting the airspace-outer space boundary according to the spatial approach and administering one set of rules on one side—domestic air law—and another set of rules on the other side—international space law—is not unsound.[114] It is far from being inconsistent and ambiguous, as has been argued, and offers a clear-cut system.

In addition, as has been explained, the breadth and delimitation of the territorial sea is not the consequence of any physical line in the sea, but the result of a chosen length added to a determined baseline. First, the breadth of 12 nm could be copied in the sense that the space boundary need not be set at any particular height. The number

---

[108] *Anglo-Norwegian Fisheries Case*, Judgment, ICJ Reports 1951, p 133, 143; James Kraska, *Maritime Power and the Law of the Sea: Expeditionary Operations in World Politics* (OUP 2011), 110; Yoshifumi Tanaka, *The International Law of the Sea* (2nd edn, CUP 2015), 49.

[109] UNCLOS, arts 5, 7, 8, 4, 12, 15, 19; David Harris and Sandesh Sivakumaran, *Cases and Materials on International Law* (8th edn, Sweet & Maxwell 2015), 336–337; Malcolm Evans, 'Law of the Sea', in Malcolm Evans (ed), *International Law* (4th edn, OUP 2015), 655.

[110] James Kraska, *Maritime Power and the Law of the Sea: Expeditionary Operations in World Politics* (OUP 2011), 109–110; David Harris and Sandesh Sivakumaran, *Cases and Materials on International Law* (8th edn, Sweet & Maxwell 2015), 337.

[111] *Maritime Delimitation and Territorial Questions between Qatar and Bahrain* (Merits), Judgment, ICJ Reports 2001, para 212; James Kraska, *Maritime Power and the Law of the Sea: Expeditionary Operations in World Politics* (OUP 2011), 110; Coalter Lathrop, 'Baselines', in Rothwell D., et al. (eds), *The Oxford Handbook of the Law of the Sea* (OUP 2015), 74.

[112] David Harris and Sandesh Sivakumaran, *Cases and Materials on International Law* (8th edn, Sweet & Maxwell 2015), 337.

[113] James Kraska, *Maritime Power and the Law of the Sea: Expeditionary Operations in World Politics* (OUP 2011) 109; Donald Rothwell and Tim Stephens, *The International Law of the Sea* (2nd edn, Hart Edition 2016), 427.

[114] Mark J. Sundahl, 'Legal status of spacecraft', in Ram S. Jakhu and Paul Stephen Dempsey (eds), *Routledge Handbook of Space Law* (Routledge Handbooks 2016), 55.

can be arbitrarily chosen, which would also preclude scientific miscalculations.[115] Persisting to search for some kind of "golden number" or physical line that does not genuinely exist is futile and falsely complexifies the issue.[116] Also, before recognizing the 12 nm breadth, most states claimed 3 nm. The fact that a small number of states explicitly refer to a 100 km breadth should not imply that an internationally agreed upon limit should follow this practice.[117]

However, where space law should depart from the law of the sea is with regard to the fact that 12 nm is the maximum. It is left to each coastal state to either claim a territorial sea up to this extent or not. By contrast, the space boundary should be settled at a universal height. It should be definitive and not open to unilateral decisions as is the case today.[118]

Second, the system of baselines should equally be omitted in the case of outer space. While it may be true that it offers flexibility and accounts for the particularities of coastlines, it is not desirable to transpose this system to the delimitation of outer space. Whether the territory of states have more mountains or more depressions should not matter. There is no unfairness in one state having a larger surface area than another, nor is there any in a state having a larger volume of airspace. Moreover, the number of abuses with regard to maritime baselines and their calculation should be an incentive for space law to leave no margin for interpretation. Although it is probable that states will not readily consent to such a proposition, ultimately it should be presented not as diminishing their sovereignty[119] but as expanding the area governed by the freedom of space exploration.[120]

For these reasons, the baseline for the outer space boundary should be set at sea level. This provides for a fictional horizontal baseline and an outer limit exactly parallel to it, which will offer clarity and certainty. No matter the location of an object around the globe, under the boundary it will be subject to one set of rules and over that altitude, to another. Let us recall that boats do not stop being boats when they

---

[115] Olavo de Oliveira Bittencourt Neto, 'Delimitation of outer space and Earth orbits', in Yanal Abul Failat and Anél Ferreira-Snyman (eds), *Outer Space Law: Legal Policy and Practice* (Globe Law and Business Ltd 2017), 59; Francis Lyall and Paul Larsen, *Space Law: A Treatise* (2nd edn, Taylor & Francis 2017), 148.

[116] S. Mishra, T. Pavlasek, 'On the Lack of Physical Bases for Defining a Boundary between Air and Space and Outer Space' (1982) 7 Annals of Air & Space Law 399, 412–413; *contra* Lubos Perek, 'Scientific Criteria for the Delimitation of Outer Space' (1977) 5 *Journal of Space Law* 111.

[117] *Contra* Olavo de Oliveira Bittencourt Neto, 'Delimitation of outer space and Earth orbits', in Yanal Abul Failat and Anél Ferreira-Snyman (eds), *Outer Space Law: Legal Policy and Practice* (Globe Law and Business Ltd 2017), 59.

[118] Olavo de Oliveira Bittencourt Neto, 'Delimitation of outer space and Earth orbits', in Yanal Abul Failat and Anél Ferreira-Snyman (eds), *Outer Space Law: Legal Policy and Practice* (Globe Law and Business Ltd 2017), 56, 59.

[119] Ibid., 53.

[120] Michael S. Dodge, 'Sovereignty and the Delimitation of Airspace: A Philosophical and Historical Survey Supported by the Resources of the Andrew G. Haley Archive' (2009) 35 *Journal of Space Law* 5, 23; Dean N Reinhardt, 'The Vertical Limit of State Sovereignty' (2007) 72 *Journal of Air Law and Commerce* 65, 119; Theodore W. Goodman, 'To the End of the Earth: A Study of the Boundary between Earth and Space' (2010) 36 *Journal of Space Law* 87, 94.

cross a zone, the nature of the object does not suddenly change because the object passes from one location to the other. The boat remains what it is, but is subjected to different regimes depending on where it sails.

Lastly, regardless of how the boundary is fixed, space law will not be able to efficiently evolve if it does not take a holistic approach and tackle other issues in parallel. Indeed, once an object has reached outer space, it will be under the jurisdiction and control of the state of registry. Depending on the level of permissiveness of the regime provided by individual states or their negligence in ensuring compliance with international standards, private companies may be more or less inclined to turn to flags of convenience. This is the object of the next section.

3.2 Flags of Convenience

"Flags of convenience", also referred to as "open registries", is a notion born out of the maritime context.[121] Boczek captured its meaning as referring to the flag of a country that allows the registration of vessels owned or controlled by foreign persons under particularly favorable conditions for the registrees.[122] In doing so, flag states often ignore their international duties and enable shipowners to benefit from more relaxed or nonexistent safety, labor, taxation, or environmental regulations.[123] The use of flags of convenience has already caused major environmental and safety incidents at sea, which are now extremely difficult to prevent and eradicate.[124]

In a similar way, the escalation in launches linked with the rise of mega constellations has the potential to cause, or at least accelerate, the phenomenon of flags of convenience in space.

Mega constellations do not only entail more launched satellites, they also mean that increased and diversified means of launching are spreading, that companies developing both constellations and launchers are blooming all over the globe, and that the space economy is growing (i.e. more money is to be made and even more

---

[121] Frans G. von der Dunk, 'Towards 'Flags of Convenience' in Space?' (2012) *IISL-ECSL Symposium* 1, 1.

[122] Boleslaw Adam Boczek, 'Flags of Convenience: An International Study' (Harvard University Press, CUP 1962), in Natalie Klein, *Maritime Security and the Law of the Sea* (OUP 2013), 15.

[123] Natalie Klein, *Maritime Security and the Law of the Sea* (OUP 2013), 15, 64; Yoshifumi Tanaka, *The International Law of the Sea* (2nd edn, CUP 2015), 162; Adrian Taghdiri, 'Flags of Convenience and the Commercial Space Flight Industry: The Inadequacy of Current International Law to Address the Opportune Registration of Space Vehicles in Flag States' (2013) 19 *Boston University Journal of Science and Technology* 405, 406, 417–418.

[124] Adrian Taghdiri, 'Flags of Convenience and the Commercial Space Flight Industry: The Inadequacy of Current International Law to Address the Opportune Registration of Space Vehicles in Flag States' (2013) 19 *Boston University Journal of Science and Technology* 405, 406.

actors attracted into the space market). All those ingredients combined together will reinforce the flaws of registration and enforcement mechanisms.[125]

The competition in the space industry is fiercer by the day, and every company dreams of reproducing the SpaceX model. To attract space business, states may elect to lower their standards—notably in terms of safety or sustainability—to attract foreign private actors, thereby creating the possibility for the development of flags of convenience in space.[126] In the case of constellations, certain states may authorize the launch of more massive constellations or constellations with concerning failure rates predictions. For example, the FCC has authorized SpaceX to launch thousands of systems despite indignation from astronomers and observed failure rates. The lawfulness of the decision has even been questioned.[127] According to reports from November 2020, Starlink's failure rate dropped from 3 to 2.5%, which is still alarming when taking into account that SpaceX is planning to launch 40,000 more systems.[128] As told by McDowell, the concern is not so much in the failure rate itself but the rather in the fact that even a "normal failure rate in such a huge constellation is going to [generate] a lot of bad space junk".[129]

3.2.1 Potentiality for Flags of Convenience in Outer Space

The legal framework for this topic is found in the Article VIII of the OST and in the Registration Convention (REG). First, the OST grants the state of registry jurisdiction and control over objects launched under its registry and over the personnel involved while in outer space.[130] As will be demonstrated below, this approach is strongly reminiscent of the exclusive jurisdiction of the flag state over vessels flying its flag while on the high seas. Second, REG requires launching states to register any launched space objects but leaves the content and conditions of the maintenance of

---

[125] Adrian Taghdiri, 'Flags of Convenience and the Commercial Space Flight Industry: The Inadequacy of Current International Law to Address the Opportune Registration of Space Vehicles in Flag States' (2013) 19 *Boston University Journal of Science and Technology* 405, 405; Matthew J. Kleiman, 'Space Law 101: An Introduction to Space Law' (American Bar 27 August 2013) www.americanbar.org/groups/young_lawyers/publications/the_101_201_practice_series/space_law_101_an_introduction_to_space_law/.

[126] Adrian Taghdiri, 'Flags of Convenience and the Commercial Space Flight Industry: The Inadequacy of Current International Law to Address the Opportune Registration of Space Vehicles in Flag States' (2013) 19 *Boston University Journal of Science and Technology* 405, 405–406, 417; Matthew J. Kleiman, 'Patent rights and flags of convenience in outer space' (2011) 23(3) *Air & Space Lawyer* 4, 5.

[127] Jonathan O'Callaghan, 'The FCC's Approval of SpaceX's Starlink Mega Constellation May Have Been Unlawful' (*Scientific American*, 16 January 2020) www.scientificamerican.com/article/the-fccs-approval-of-spacexs-starlink-mega-constellation-may-have-been-unlawful/.

[128] Morgan McFall-Johnsen, 'About 1 in 40 of SpaceX's Starlink satellites may have failed' *Business Insider* (3 November 2020) www.businessinsider.com/spacex-starlink-internet-satellites-percent-failure-rate-space-debris-risk-2020-10?r=DE&IR=T.

[129] Ibid.

[130] Outer Space Treaty, art VIII.

such registry to the discretion of the state of registry.[131] REG also addresses what information the state of registry must submit, such as the name of the launching state and other technical details.[132] Surprisingly, no disposition seems to mimic the genuine link requirement between a ship and a flag state present in the law of the sea.[133] Further, there is no regulation for the certification of space technologies, the crew, or the general safety.[134] These aspects may provide an incentive to states of registry to compete with each other using appealing regulations for private companies.[135]

Although this does not provide much clarification, the "state of registry" is defined as a launching state on whose registry a space object is registered in accordance with Article II of REG, whereas the "launching state" shall refer to either the state that launches or procures the launching of a space object, or the state from whose territory or facility such a launch is operated.[136] This broad definition of the launching state effectively gives space actors a choice to subject themselves to the most advantageous jurisdiction, to the state that offers the "best deal".[137]

Yet, in spite of these registration regulations, disputes and incidents have effectively been dealt with according to liability provisions.[138] On the one hand, the OST declares states internationally responsible for national activities whether conducted by public or private actors and for ensuring their conformity with the Treaty.[139] On the other hand, the OST holds launching states internationally liable for damages ensuing from the participation in the launch of a space object.[140] Interestingly, this approach diverges from the sea regime, in terms of which states have responsibility and liability obligations towards their own conduct and shipowners are held privately liable for damage caused by their ships.[141]

---

[131] Registration Convention, arts II(1), II(3).

[132] Registration Convention, art IV.

[133] Adrian Taghdiri, 'Flags of Convenience and the Commercial Space Flight Industry: The Inadequacy of Current International Law to Address the Opportune Registration of Space Vehicles in Flag States' (2013) 19 *Boston University Journal of Science and Technology* 405, 406; Frans G. von der Dunk, 'Towards 'Flags of Convenience' in Space?' (2012) *IISL-ECSL Symposium* 1, 5.

[134] Frans G. von der Dunk, 'Towards 'Flags of Convenience' in Space?' (2012) *IISL-ECSL Symposium* 1, 4.

[135] Adrian Taghdiri, 'Flags of Convenience and the Commercial Space Flight Industry: The Inadequacy of Current International Law to Address the Opportune Registration of Space Vehicles in Flag States' (2013) 19 *Boston University Journal of Science and Technology* 405, 406.

[136] Registration Convention, art I.

[137] Adrian Taghdiri, 'Flags of Convenience and the Commercial Space Flight Industry: The Inadequacy of Current International Law to Address the Opportune Registration of Space Vehicles in Flag States' (2013) 19 *Boston University Journal of Science and Technology* 405, 419.

[138] Frans G. von der Dunk, 'Towards 'Flags of Convenience' in Space?' (2012) *IISL-ECSL Symposium* 1, 5.

[139] Outer Space Treaty, art VI.

[140] Outer Space Treaty, art VII, Liability Convention, art II.

[141] Outer Space Treaty, arts VI, VII; Armel Kerrest, 'Space law and the law of the sea', in Christian Brünner and Alexander Soucek (eds), *Outer Space in Society, Politics and Law* (Springer Wien New York 2011), 254; Adrian Taghdiri, 'Flags of Convenience and the Commercial Space Flight

Surely, absolute liability may hinder states in enabling reckless behavior by private actors.[142] However, enforcement procedures in this field are largely unreliable, if they exist at all, and states acting with disregard for the law, even if facing punitive actions, cannot be compelled to comply.[143] There are no tangible negative consequences to this sort of abuse.

Besides, states of registry can enact in their domestic legislation third-party liability provisions to recoup what was paid on account of the *de facto* responsible private actor. Nothing prevents states from doing so while giving advantage to private actors. Depending on the substance of the arrangement, whether the state imposes insurance requirements or not on its private actors, or depending on the threshold of the coverage, companies would be inclined to choose one state over another.[144] In reality, varying and relatively diverging third-party liability laws have been deployed by launching states.[145]

Overall, this *modus operandi* seems highly counterintuitive and counterproductive, in that triggering the liability regime assumes the problem has already occurred.[146] Thus, instead of dealing with the issue at its roots, before or during registration, states choose to stand by during an incident for which they are officially responsible and liable. Then, subsequently, they have recourse to their domestic third-party liability process, if any exists.

All these potential issues are exacerbated by the fact that REG is no longer in touch with today's technological realities. Not only have new activities been introduced and many new nations now have space capabilities and programs, multiplying the launching of objects from an array of possible territories, but also the involvement of private actors has exploded.[147] For example, around 190 satellites constellations

---

Industry: The Inadequacy of Current International Law to Address the Opportune Registration of Space Vehicles in Flag States' (2013) 19 *Boston University Journal of Science and Technology* 405, 405, 423; Frans G. von der Dunk, 'Towards 'Flags of Convenience' in Space?' (2012) *IISL-ECSL Symposium* 1, 5–6.

[142] Adrian Taghdiri, 'Flags of Convenience and the Commercial Space Flight Industry: The Inadequacy of Current International Law to Address the Opportune Registration of Space Vehicles in Flag States' (2013) 19 *Boston University Journal of Science and Technology* 405, 406; Armel Kerrest, 'Space law and the law of the sea', in Christian Brünner and Alexander Soucek, *Outer Space in Society, Politics and Law* (2011, Springer Wien New York), 249–250; Frans G. von der Dunk, 'Towards 'Flags of Convenience' in Space?' (2012) *IISL-ECSL Symposium* 1, 14.

[143] Adrian Taghdiri, 'Flags of Convenience and the Commercial Space Flight Industry: The Inadequacy of Current International Law to Address the Opportune Registration of Space Vehicles in Flag States' (2013) 19 *Boston University Journal of Science and Technology* 405, 405, 423.

[144] Frans G. von der Dunk, 'Towards 'Flags of Convenience' in Space?' (2012) *IISL-ECSL Symposium* 1, 5–6, 13.

[145] Ibid., 1, 13.

[146] Ibid., 1, 14–15.

[147] Frans G. von der Dunk, 'Towards 'Flags of Convenience' in Space?' (2012) *IISL-ECSL Symposium* 1, 14–15; United Nations General Assembly, 62nd session, Resolution Adopted by the General Assembly, 'Recommendations on enhancing the practice of States and international intergovernmental organizations in registering space objects', A/RES/62/101 (10 January 2008) www.unoosa.org/pdf/gares/ARES_62_101E.pdf, 2.

from private actors are currently at different stages of development, whilst around 110 micro launchers are at development or conception stages.[148]

This in turn has increased the risks of incidents and disputes, and the likelihood of environmental, labor or safety standards being relaxed by states of registry in order to attract companies.[149] The fact that the FCC is authorizing SpaceX to launch an immense number of satellites despite the results of studies evaluating failure rates and concerns for astronomy may draw many companies aspiring to launch small satellites constellations on the U.S. market.

Lastly, in the scenario where a state has suffered damage from an unidentified space object, or if the unidentified object is potentially dangerous, the sole remedy provided to states under REG is to allow the state in question to seek assistance from other states in the identification process and to ask those states to respond to the best of their capabilities.[150] Evidently, this procedure rests on the good faith of other states. However, in the context of flags of convenience, not much good faith can be expected. As summarized by Taghdiri, the current legal landscape has been shaped by international agreements but remains mostly silent on enforcement measures in the case of breaches of international standards and disputes.[151] Consequently, facing no serious repercussions, states may once more be tempted to disregard international standards, relax their domestic regulations with the aim of attracting foreign private actors, and hide their identity from interested states.[152]

3.2.2 Flags of Convenience at Sea

Save for a few exceptions,[153] it emerges from the Permanent Court of International Justice in the *SS Lotus* case and Article 92 of the UNCLOS that flag states are the sole bearers of authority over ships flying their flags on the high seas.[154] This principle aims to keep the high seas immune to national sovereignty and puts the flag state in charge of the supervision and legal compliance of its vessels.[155] Somewhat shielded by this exclusive jurisdiction, some flag states decide to forsake maritime security and instead prioritize their own interests.[156]

---

[148] Eric Kulu, 'NewSpace Index' newspace.im.

[149] Frans G. von der Dunk, 'Towards 'Flags of Convenience' in Space?' (2012) *IISL-ECSL Symposium* 1, 14–15.

[150] Registration Convention, art VI.

[151] Adrian Taghdiri, 'Flags of Convenience and the Commercial Space Flight Industry: The Inadequacy of Current International Law to Address the Opportune Registration of Space Vehicles in Flag States' (2013) 19 *Boston University Journal of Science and Technology* 405, 415.

[152] Ibid., 405, 406.

[153] UNCLOS, arts 110, 111.

[154] *SS Lotus Case (France v Turkey)* (1927) PCIJ, Serie A, No. 10, p 25; UNCLOS, art 92; Natalie Klein, *Maritime Security and the Law of the Sea* (OUP 2013), 106–107.

[155] Natalie Klein, *Maritime Security and the Law of the Sea* (OUP 2013), 106–107.

[156] Ibid., 64.

In order to attract foreign companies for their own economic enrichment, certain states are willing to disregard environmental or safety standards, to offer distinctly competitive financial and taxation arrangements, as well as more relaxed labor laws.[157]

Furthermore, Klein has pointed out another problem created by flag states' irresponsible behavior. When a ship breaches international law on the high seas, according to Article 94(6), a foreign state has the option to report the unlawful activities to the flag state, which should then take matters into its figurative hands.[158] However, if that state is unwilling or unable to investigate or prosecute the wrongdoer, or if the latter suddenly changes its state of registry, the whole system crashes.[159] Exactly as noted in the previous section regarding the remedy offered by Article VI of REG, the core of this enforcement system reveals deep flaws, in that the flag or registry state is judge, jury and wrongdoer.

The effects of flags of convenience at sea are varied. Most notably, the non-observance of international standards by shipowners, and the non-enforcement of these standards by flag states, have irreversibly impacted the environment. Indeed, to this day, flags of convenience remain behind many maritime incidents and ecological disasters.[160] Notoriously, the Deepwater Horizon incident of 2010 not only killed crew members, but also seriously contaminated the sea and marine life.[161] At the time of the oil rig's explosion in the Gulf of Mexico, and more precisely in the EEZ of the U.S., it was registered in the Marshall Islands and owned by a Swiss company.[162] It was later alleged that the registration to this flag was done in order to benefit

---

[157]Natalie Klein, *Maritime Security and the Law of the Sea* (OUP 2013), 64; Douglas Guilfoyle, 'The High Seas', in Donald Rothwell et al. (eds), *The Oxford Handbook of the Law of the Sea* (OUP 2015), 216.

[158]UNCLOS, art 94(6); Natalie Klein, *Maritime Security and the Law of the Sea* (OUP 2013), 106–107; Richard Barnes, 'Flag States', in Rothwell D., et al. (eds), *The Oxford Handbook of the Law of the Sea* (OUP 2015), 309.

[159]Natalie Klein, *Maritime Security and the Law of the Sea* (OUP 2013), 106–107.

[160]Adrian Taghdiri, 'Flags of Convenience and the Commercial Space Flight Industry: The Inadequacy of Current International Law to Address the Opportune Registration of Space Vehicles in Flag States' (2013) 19 *Boston University Journal of Science and Technology* 405, 418; Frans G. von der Dunk, 'Towards 'Flags of Convenience' in Space?' (2012) *IISL-ECSL Symposium* 1, 2.

[161]Oliver Milman, 'Deepwater Horizon disaster altered building blocks of ocean life' (The Guardian, 28 June 2018) www.theguardian.com/environment/2018/jun/28/bp-deepwater-horizon-oil-spill-report; Richard Black, 'Gulf oil leak: Biggest ever, but how bad?' (BBC News, 3 August 2010), www.bbc.co.uk/news/science-environment-10851837; Adrian Taghdiri, 'Flags of Convenience and the Commercial Space Flight Industry: The Inadequacy of Current International Law to Address the Opportune Registration of Space Vehicles in Flag States' (2013) 19 *Boston University Journal of Science and Technology* 405, 418.

[162]Richard Pallardy, 'Deepwater Horizon oil spill of 2010' (Britannica, 9 July 2010) www.britannica.com/event/Deepwater-Horizon-oil-spill-of-2010; Adrian Taghdiri, 'Flags of Convenience and the Commercial Space Flight Industry: The Inadequacy of Current International Law to Address the Opportune Registration of Space Vehicles in Flag States' (2013) 19 *Boston University Journal of Science and Technology* 405, 418.

from more relaxed safety standards and that the inspection of sites like Deepwater Horizon, which should take weeks, was accomplished in a matter of hours.[163]

Unfortunately, environmental and safety negligence is not all that flag states are willing to do in order to lure shipowners. Subjecting private companies to less strict labor laws requiring inferior qualifications and low-cost labor are also part of their tactics.[164] Once again, the motive is financial. As noted by Tanaka, crew expenses are one of the few variables in the shipping industry, and as such, a non-negligible way for ill-intentioned shipowners to make budgetary cutbacks.[165]

In addition to the use of flags of convenience for economic profits, they may also be resorted to for strategic military purposes.[166] Indeed, during a conflict, an enemy ship might use the flag of a neutral state to hide its true origin and involvement in the conflict.[167] This creates another issue: if the recourse to this stratagem is discovered, belligerent ships could be tempted to verify the neutral nature of all possible ships, which may result in widespread interferences.[168]

At some point, the law of the sea sought to resolve the issue by initiating a genuine link requirement between a ship and the state whose flag it flies.[169] Unfortunately, the requirement seriously lacks substance. Actually, ITLOS asserted in *M/V Virginia G* that mere registration is sufficient to establish the genuine link.[170] Still, this link is not a one-way street and establishing the genuine link should not stop there. While shipowners simply have to register a vessel to manifest a connection, as recalled in *M/V Virginia G* and under Article 94 of the UNCLOS flag states have a duty to exercise effective control and jurisdiction over their ships and ensure their compliance with international law.[171] Consequently, can a genuine link by definition ever be demonstrated in the face of relations between shipowners and states offering flags of convenience? If the answer is negative, will there be an authority to review the link

---

[163] Adrian Taghdiri, 'Flags of Convenience and the Commercial Space Flight Industry: The Inadequacy of Current International Law to Address the Opportune Registration of Space Vehicles in Flag States' (2013) 19 *Boston University Journal of Science and Technology* 405, 418; BBC News, 'US oil spill: 'Bad management' led to BP disaster' (BBC News, 6 January 2011) www.bbc.co.uk/news/world-us-canada-12124830.

[164] Yoshifumi Tanaka, *The International Law of the Sea* (2nd edn, CUP 2015), 162.

[165] Ibid.

[166] Natalie Klein, *Maritime Security and the Law of the Sea* (OUP 2013), 290–291.

[167] Ibid.

[168] Ibid., 291.

[169] UNCLOS, art 91(1); Natalie Klein, *Maritime Security and the Law of the Sea* (OUP 2013), 106–107; Donald Rothwell and Tim Stephens, *The International Law of the Sea* (2nd edn, Hart Edition 2016), 376–377; Frans G. von der Dunk, 'Towards 'Flags of Convenience' in Space?' (2012) *IISL-ECSL Symposium* 1, 2.

[170] *M/V "Virginia G" (Panama/Guinea-Bissau)*, Judgment, ITLOS Reports 2014, para 113; Natalie Klein, *Maritime Security and the Law of the Sea* (OUP 2013), 106–107; Donald Rothwell and Tim Stephens, *The International Law of the Sea* (2nd edn, Hart Edition 2016), 376–377.

[171] *M/V "Virginia G" (Panama/Guinea-Bissau)*, Judgment, ITLOS Reports 2014, p 4, para 113; UNCLOS, art 94; Donald Rothwell and Tim Stephens, *The International Law of the Sea* (2nd edn, Hart Edition 2016), 376–377.

and seek its true nature?[172] On this issue, another compelling observation has been made by Klein. Whereas in times of peace, the genuine link remains a mere synonym of registration, during armed conflicts, prize law—which concerns the seizure of enemy property during naval warfare—provides a review system for ships and their connection to neutral flags.[173] In that case, the focus is on the actual ownership of a vessel instead of its registration.[174]

Another formula to resolve issues associated with flags of convenience may be to erode the exclusive flag state jurisdiction. Besides the usual exceptions, such as the right of hot pursuit and the right of visit, Article 218 grants port states with additional enforcement measures in order to protect the marine environment.[175] It does so by circumventing inefficient enforcement procedures and empowering port states to investigate and prosecute any illegal discharge from vessels that voluntarily entered their port, even if it has been perpetrated outside the internal waters, the territorial sea, or the EEZ of the port state.[176] In other words, some sort of protection against flag states not fulfilling their duties and undermining the system is offered by granting additional competences to different levels of authority.

3.2.3 Preventing Flags of Convenience in Outer Space: Adapting the Successes and Failures of the Law of the Sea

As demonstrated, the law of the sea is not able to effectively deal with the flags of convenience problematic. Yet, it may offer valuable insights into what the situation could become in space and how to prevent this from happening.

First, regarding the genuine link requirement, a similar requirement could be envisioned between satellite owners and states of registry. However, this link would need to have more substance than just the evidence of registration. To begin, a stricter definition of launching state for the purpose of registration should be used, so as not to enable private actors to cherry-pick the most favorable state. Furthermore, Von der Dunk has proposed remodeling Article 91(1) of the UNCLOS for the requirements of space law. The measure would still recognize the right of states to determine their registration conditions and content, as in Article II(3) of REG, but it would require a genuine link, the existence of which should be emphasized by the exercise of jurisdiction and control by states over registered space objects, similar to what is

---

[172]Richard Barnes, 'Flag States', in Rothwell D., et al. (eds), *The Oxford Handbook of the Law of the Sea* (OUP 2015), 309.

[173]Natalie Klein, *Maritime Security and the Law of the Sea* (OUP 2013), 291; James Kraska, 'Prize Law' (2009) *Max Planck Encyclopedia of Public International Law* 1, para 1.

[174]Natalie Klein, *Maritime Security and the Law of the Sea* (OUP 2013), 291.

[175]UNCLOS, arts 110, 111; Natalie Klein, *Maritime Security and the Law of the Sea* (OUP 2013), 69–71; Donald Rothwell and Tim Stephens, *The International Law of the Sea* (2nd edn, Hart Edition 2016), 376–377.

[176]UNCLOS, art 218(1); Natalie Klein, *Maritime Security and the Law of the Sea* (OUP 2013), 69–71; Donald Rothwell and Tim Stephens, *The International Law of the Sea* (2nd edn, Hart Edition 2016), 376–377.

found in Article 94(1) of the UNCLOS.[177] Further to this idea, it could also include, as in Article 94, paragraphs 3 to 5, a list of safety measures to be enforced by the state of registry with respect to things such as the construction of space objects, their surveying and manning, or the qualifications of persons involved.

In contrast, Article 94(6) is as impractical as Article VI of REG. Thus, in order to supplant the authority of the state of registry when a flag of convenience situation is suspected, a review process such as the one established in prize law could be imagined. As such, the nationality of the object owner could surpass the state of registry. However, the question shifts towards what authority would take that role. Kleiman has suggested, in the context of the effect of flags of convenience on patent protection, an idea which can be used more generally. Essentially, he has introduced the possibility of a multinational jurisdiction.[178] Like with ITLOS in the *M/V Saiga* (No. 2) and *Grand Prince* cases, one could imagine an international court or tribunal having as a competence the verification of the genuine link with regard to space activities.[179]

If the preceding paragraph suggests a clarification of international norms, a second lesson would point towards the harmonization of domestic ones.[180] In 2008, the UN General Assembly made various recommendations regarding state practice in registering space objects.[181] It underlined the changed nature of the space industry engendered by the multiplication of new technologies and new actors, governmental as much as private.[182] In consequence, it vouched for the harmonization of national laws and proposed the submission by states to the UN of a number of additional pieces of information.[183] The issue of flags of convenience at sea is extremely tenacious and almost impossible to eradicate, precisely because of the great disparities between national regimes. If harmony is to be reached in space law, it has to be achieved now that great divergences between states have not yet appeared and incidents due to flags of convenience have not yet manifested as such.[184] Cooperation can still be pursued while states are mostly compliant and of good faith, and while the commercial space

---

[177] Frans G. von der Dunk, 'Towards 'Flags of Convenience' in Space?' (2012) *IISL-ECSL Symposium* 1, 16.

[178] Matthew J. Kleiman, 'Patent rights and flags of convenience in outer space' (2011) 23(3) *Air & Space Lawyer* 4, 6.

[179] *"Grand Prince" (Belize v. France)*, Prompt Release, Judgment, ITLOS Reports 2001, paras 81–93; *M/V "Saiga" (No 2) (Saint Vincent and the Grenadines v Guinea)*, Judgment, ITLOS Reports 1999, paras 75–87; Douglas Guilfoyle, 'The High Seas', in Donald Rothwell et al. (eds), *The Oxford Handbook of the Law of the Sea* (OUP 2015), 209.

[180] Frans G. von der Dunk, 'Towards 'Flags of Convenience' in Space?' (2012) *IISL-ECSL Symposium* 1, 16.

[181] United Nations General Assembly, 62nd session, Resolution Adopted by the General Assembly, 'Recommendations on enhancing the practice of States and international intergovernmental organizations in registering space objects', A/RES/62/101 (10 January 2008) www.unoosa.org/pdf/gares/ARES_62_101E.pdf.

[182] Ibid., 2.

[183] Ibid., 3(2).

[184] Frans G. von der Dunk, 'Towards 'Flags of Convenience' in Space?' (2012) *IISL-ECSL Symposium* 1, 15–16.

industry is still in its infancy.[185] The 1986 UN Registration Convention seeking to reevaluate the genuine link which has not yet entered into force is a good example that once states defend profoundly disparate interests, it will be too late to reach a consensus.[186] Also, just like with the law of the sea, the current legal loopholes, voids, and disagreements will inevitably worsen the situation once the damage is done.[187]

Third, it has been noted that flags of convenience have appeared at the registration level, but also concern liability norms. As mentioned previously, the harmonization of domestic procedures is key. However, the international space regime seems to have an advantage over the regime at sea. Indeed, where the law of the sea holds private actors accountable, states are internationally responsible for national activities and liable for damage. This ensures the ability of victims to sue the "bigger fish", and therefore to access the financial resources a private company may not have. Unfortunately, it is to be expected, as cautioned by Von der Dunk, that once the occurrence of damage caused by space activities increases, it could result in enormous claims that states might not be able to sustain.[188] States would also have minimal chances to be reimbursed by the *de facto* party responsible, the private company.[189] Yet, harmonization of domestic laws could deal with this issue and it seems particularly prudent not to copy the maritime model in this matter, but to maintain the international responsibility and liability of the states.

Fourth, the idea to give additional authority to port states seems hardly transposable to the context of space activities. Perhaps it could be envisioned if the notion of the launching state was constrained to one possible state. In other words, one could imagine the launching state defined as the state of the entity providing launch services, the one operating the launch. Then, the state from whose territory the space object is launched, could constitute a sort of "ground state". The two could be the same or different. For instance, when Arianespace procures launch services through a Soyuz launcher at the Baikonur spaceport in Russia, France is the launching state, and Russia could be the ground state.

Eventually, if nothing is done, it is to be expected that the development of new activities, not yet regulated, will certainly lead to flags of convenience in outer space. That is especially to be feared from commercial activities, such as the exploitation of extraterrestrial resources, where countries such as Luxembourg attract dozens of foreign companies because of their liberal regime.[190]

---

[185] Adrian Taghdiri, 'Flags of Convenience and the Commercial Space Flight Industry: The Inadequacy of Current International Law to Address the Opportune Registration of Space Vehicles in Flag States' (2013) 19 *Boston University Journal of Science and Technology* 405, 419–420.
[186] Yoshifumi Tanaka, *The International Law of the Sea* (2nd edn, CUP 2015), 163.
[187] Frans G. von der Dunk, 'Towards 'Flags of Convenience' in Space?' (2012) *IISL-ECSL Symposium* 1, 14–15.
[188] Ibid., 1, 14.
[189] Ibid., 1, 14.
[190] Justin Calderon, 'The tiny nation leading a new space race' (BBC News, 16 July 2918) www.bbc.com/future/story/20180716-the-tiny-nation-leading-a-new-space-race.

4 Conclusion

Mega constellations should compel law-makers and more generally decisions-takers to come together and agree on ways to address new trends, such as mega constellations, and legally frame the effects of those activities. They do affect the entire planet and everyone is concerned; each state should have their word to say and each state should take its responsibilities. It has to be done now that the worst has not yet happened. Is it not in the nature of lawyers to foresee potential problems before they occur so that once they do, the attention is centralized on finding solutions and moving forward?

Against this backdrop, space law can be guided by the international law of the sea which has developed over centuries. Sometimes it has been innovative, while at other times it was not able to foresee future difficulties and failed to provide solutions to major problems. In this light, it can be a great source of inspiration and influence for international space law, whether through learning from its strengths or its flaws.

The first analogy explored was on the delimitation of boundaries in outer space and at sea. It was demonstrated that as long as the airspace-outer space limit is not settled, and outer space is not properly defined, space law will never truly be coherent and clear. Indeed, if one does not know where outer space begins, one cannot possibly know where the national territories and sovereignty of states end. One cannot distinguish where international territories begin. Contrary to the former, the latter is not subject to assertion of sovereignty and only states of registry have jurisdiction and control over the objects registered under their flags. This difference between the legal regimes in air space and outer space makes it crucial to establish the extent of each jurisdiction. The analysis also revealed that none of the existing approaches to setting that limit are ideal. Consequently, the breadth and delimitation of the territorial sea were investigated, since its regime is in opposition with that of the high seas. It was then concluded that if space law may use the same logic that led to the establishment of the 12 nm, the airspace-outer space limit should be universal and based on sea level.

The second analogy reviewed was on flags of convenience and the probability of their occurrence in space activities, considering their proliferation at sea. The flaws that exist in present space regimes, whether it be in regard to registration, liability or enforcement, will be intensified in view of the growth of the space industry that is to come. The necessity for action was demonstrated and it was explained that such action could be modeled after different provisions of the law of the sea, such as the genuine link requirement and the powers given to certain levels of authority to supplant the exclusive jurisdiction of the flag state in particular circumstances. However, a way should be found to improve these provisions, and this should be done before the problem actually occurs in outer space. A possible solution would be to give an authority the competence to review suspicious claims of genuine link and by redefining the concept of "launching state".

Ultimately, the article sought to demonstrate that, contrary to certain beliefs, the law of the sea is apt to help space law get out of the stagnation it has been in for many years. The law of the sea has encountered many problems. It has faced some with great success and it has failed in respect of others. However, in both regards, it can serve as a lesson for outer space. Space law should not seek to copy the law of the sea. It should use the available experiences to tailor a regime that accounts for each of its particularities, while anticipating potential obstacles, shape the way forward, and envision a reality that contains such things as mega constellations.

Lauryn Hallet holds an LL.M in International Law from the University of Bristol and is a researcher at the European Institute of Space Policy (ESPI). Her research areas include the use of force, maritime security, and relations between outer space and the sea, as well as their respective legal frameworks. It also includes Space Law in the context of the exploitation of natural resources, delimitation of outer space, flags of convenience, and human exploration of Mars. She is currently working on regional space developments in Latin American and Caribbean countries.

Printed in Great Britain
by Amazon